Environmental good practice on site

Stuart Coventry
Claire Woolveridge

CIRIA

sharing knowledge
building best practice

6 Storey's Gate, Westminster, London SW1P 3AU
Telephone 020 7222 8891 Fax 020 7222 1708
Email enquiries@ciria.org.uk
Website www.ciria.org.uk

Summary

Construction sites are often criticised for the damage that they cause to the surrounding environment and the adverse effects that they have on their neighbours. This can take many forms, for example, effects on the natural environment and watercourses, excessive noise and pollution of the air. Good practice on site to preserve our environment is now usually a high priority for clients, their professional advisers, contractors and regulators.

This handbook provides practical guidance for site managers, site engineers and supervisors on how to manage construction on site to control environmental impacts. The handbook has four chapters:

- **Introduction.** An outline of the benefits of good practice and the environmental obligations under which a site operates in terms of both the legislation and the contract conditions.

- **General site management issues.** This chapter provides general guidance on good site practice to provide the framework for managing environmental impacts. Common procedures are covered, for example: establishing the management framework, ensuring good public relations, securing the site and managing materials.

- **Environmental issues.** For each environmental issue, from water to archaeology, detailed guidance is provided on how to prevent impacts, and how to recognise and deal with any problems that may arise.

- **Construction processes.** This chapter identifies the environmental issues that need to be considered when carrying out a particular construction process, for example, piling.

The handbook is intended as a user-friendly guide, a reference book and a training aid. It is accompanied by a poster (C502P) and by a pocket book (C503) that present the key advice to construction workers.

Environmental good practice on site

Construction Industry Research and Information Association

This edition first published 1999. Reprinted 2000, 2001 and 2002

C502 © CIRIA 1999   ISBN 0 86017 502 2

Keywords:	archaeology, construction site, contamination, dust, ecology, emissions, environment, management, noise, odours, vibration, waste, water.

Reader interest	**Classification**	
contractors, site workers, site managers, engineers, supervisors, architects	Availability	Unrestricted
	Content	Advice/guidance
	Status	Committee-guided
	Users	Construction professionals and managers

Acknowledgements

This is the project report for CIRIA Research Project RP559, "Site guide – environmental good practice". This publication also constitutes Environment Agency R&D Publication 15.

The work was carried out by Scott Wilson with support from Balfour Beatty. The principal contributors from Scott Wilson were Stuart Coventry and Claire Woolveridge, and from Balfour Beatty, Marcus Pearson. The handbook was designed by Thirst Design and Marketing.

Funders

The research leading to the publication of this site guide was funded by CIRIA Core Programme, the Environment Agency, Sniffer (Scotland & Northern Ireland Forum for Environmental Research) and the National House Building Council.

Project steering group

CIRIA wishes to express its thanks to the members of the project steering group for their contributions to the work:

Andrew Beauchamp (chairman)	*Connect Roads Ltd*
Phil Chatfield	*Environment Agency*
Trevor Curson	*Aspinwall & Co Ltd*
Lawrence Emmerson	*BAA*
Rob Kremis	*Tarmac*
Arthur Perks	*National Housebuilding Council*
Gerard Sloyan	*J Doyle & Co Demolition Ltd*
Brent Turton	*Westminster City Council*
Martin Worthington	*AMEC Civil Engineering.*

Thanks also go to the corresponding members of the project steering group:

Kelvin Potter	*ICI Technology*
Ruth Wolstenholme	*Scottish Environment Protection Agency*
Richard Wright	*English Nature.*

CIRIA's research manager for this project was Daniel J Leggett.

Source materials

CIRIA is grateful to the following organisations which provided information and access to sites:

AOC Archaeology
Costain Civil Engineering Ltd
Rydon Construction
Schal
Tarmac.

Contents

Glossary 6
The target audience. 7
How to use this handbook 8
Coverage of this handbook 9
References to the Environmental Agencies 10
Relationship to other CIRIA guidance. 11

1 Benefits and obligations 13
1.1 Introduction. 14
1.2 The benefits of good practice 15
1.3 Environmental obligations 16

2 General site management issues 17
2.1 The management framework. 18
2.2 Setting up and managing the site 26

3 Environmental issues. 35
3.1 Water 36
3.2 Waste 48
3.3 Noise and vibration 57
3.4 Dust, emissions and odours. 72
3.5 Ground contamination 80
3.6 Wildlife and natural features 85
3.7 Archaeology 94

4 Construction processes 99
4.1 Introduction. 100
4.2 Bored tunnelling 100
4.3 Brick/blockwork. 101
4.4 Concrete batching 101
4.5 Concrete pours and aftercare 102
4.6 Demolition. 103
4.7 Dredging 104
4.8 Earthworks. 105
4.9 Excavation 106
4.10 Grouting. 108
4.11 Microtunnelling. 108
4.12 Piling (including temporary works). 109
4.13 Plant maintenance 110
4.14 Refurbishment of buildings 110
4.15 Repairs to exposed structural elements (eg bridge soffits, cladding etc) 111
4.16 Roadworks. 111
4.17 Temporary works. 112
4.18 Use of oils and chemicals 112
4.19 Use of small plant 115
4.20 Working near water 116
4.21 Working with groundwater 119

Glossary

Archaeology	The study of historic remains, often by excavation.
Controlled waters	Virtually all natural waters in the UK. It includes rivers, streams, ditches, ponds and groundwater. Responsibility for policing controlled waters is placed with the Environmental Agencies.
Dust	Airborne solid matter up to about 2 mm in size.
Ecology	All living things, such as trees, flowering plants, insects, birds and mammals, and the habitats in which they live.
Environment	Both the natural environment (air, land, water resources, plant and animal life), and the human environment.
Environmental Agencies	These include the Environment Agency, the Scottish Environment Protection Agency and the Northern Ireland Environment and Heritage Service. See the References to the Environmental Agencies section on page 10 for more details.
Heritage bodies	These have a general duty to conserve our heritage, to carry out scheduling of historic remains, and to undertake research. They comprise English Heritage, Cadw, Historic Scotland, and the Northern Ireland Environment and Heritage Service.
Nature conservation bodies	The four organisations that have regional responsibility for promoting the conservation of wildlife and natural features: Countryside Council for Wales, English Nature, Northern Ireland Environment and Heritage Service, and Scottish Natural Heritage.
Noise	Often explained as being a sound that is not desired. Sound is a wave motion carried by air particles between the source and the receiver, usually the ear.
Pollution	The introduction of a substance that has the potential to cause harm to the environment. Pollutants include silty water, oils, chemicals, litter and mud.
Recycling	Collecting and separating materials from waste and processing them to produce marketable products.
Reduction	Waste reduction has two components: ● reducing the amount of waste produced ● reducing the hazard of the waste produced.
Regulatory agencies	These agencies have responsibility for enforcing environmental legislation. Refer to page 10 for more information.
Reuse	Putting objects back into use, without processing, so that they do not remain in the waste stream.
Waste	Any substance or object that the holder discards, intends to discard, or is required to discard.

The target audience

Environmental issues arise throughout a construction project. People working in construction have to be aware of their environmental obligations and the benefits that good practice will bring at every stage from the initial feasibility studies through to design, construction planning and the actual works on site. As the environmental issues differ at each stage, the approach to resolve them may also differ accordingly.

This handbook provides guidance on environmental good practice for the construction stage. It is therefore primarily aimed at site staff:

- site managers
- site engineers
- site foremen and site supervisors
- project managers
- contract supervisors/resident engineers.

It is relevant to all organisations represented on a construction site, whether as a main contractor or a sub-contractor.

The consequences of environmental planning, or lack of planning, by people involved in the early stages of a project's development can have a profound effect on the ability of the site staff to meet their obligations. Due to this, other construction professionals should seek to understand the site environmental good practice presented in this handbook. These include:

- construction planners within contractors' main offices
- contractors
- project managers/directors
- designers
- clients
- construction managers
- planning supervisors and principal contractors.

Much of the advice contained in this handbook is based on practices that have been carried out on construction sites for many years, but many of the ideas are recent. It is therefore a handbook for all levels of construction experience, from the young site engineer to the experienced site manager.

How to use this handbook

This handbook is written for a wide range of construction personnel. Chapter 1 presents the benefits of reading this handbook and the reasons for adopting good environmental practice on site.

Good environmental practice starts with good site management. Chapter 2 shows how the overall establishment and management of the site can form the basis of environmental good practice. Therefore those with a management responsibility, from the beginning of the set-up of the site right through to site completion and clear-up, should read this chapter.

If you want to know about a particular environmental issue (eg noise or archaeology), then turn to Chapter 3 for guidance. If you need to know what environmental issues to consider for a particular construction activity (eg piling), then turn to that activity in Chapter 4.

The handbook is intended as a user-friendly guide, a reference book and a training aid. It is accompanied by a poster and by a pocket book that present the key advice to construction workers. Use this handbook in toolbox talks by concentrating on the issue or construction activity most relevant to the work taking place at that time.

Throughout the handbook, symbols alongside the text help you to identify the type of information that is presented. These symbols and their explanations are shown below.

Case study

Court case

Plan ahead

Checklist

Emergency

Key guidance

Coverage of this handbook

This handbook addresses environmental issues once a project has reached the construction stage. For each of the key topics it describes the issues and how to avoid or overcome them. In order to facilitate good environmental practice on site, this handbook explains how to establish a suitable management framework for the site, and how to set up central site facilities such as vehicle refuelling stations and materials storage locations.

The reader should be clear about the limitations of scope of the handbook. In particular:

- it is not a health and safety manual
- it should not replace contact with regulators
- although it gives an overview of legislation, detailed guidance should be sought from the company's environmental representative (or external specialists) if it is required
- in all instances when dealing with the issues covered do not take action beyond your expertise. If in doubt, seek specialist advice
- it does not cover the environmental issues that should be covered during the planning and design of the project.

The guidance is generally relevant for all types of contract conditions, eg traditional, design and build, design build finance operate. However, the site manager should be aware that on some types of contract the contractor carries the risk of cost or programme delays caused by unexpected events or finds. If in doubt, seek guidance from your company's environmental representative.

References to the Environmental Agencies

When a reference is made to the Environmental Agencies it should be read to include:

- **Environment Agency (with jurisdiction over England and Wales)**
- **Scottish Environment Protection Agency (SEPA)**
- **Northern Ireland Environment and Heritage Service.**

Obtain the contact numbers for the Environmental Agencies from Directory Enquiries.

CIRIA has produced three key publications that establish the environmental issues to be addressed at all stages of construction:

- *A clients' guide to greener construction,* CIRIA Special Publication 120
- *Environmental handbook for building and civil and engineering projects: design phase,* CIRIA Special Publication 97
- *Environmental handbook for building and civil and engineering projects: construction phase,* CIRIA Special Publication 98.

These key publications outline the issues, principles and legislation that should be adopted to improve environmental performance in the construction industry. Other CIRIA publications of particular relevance are:

- *Building a cleaner future,* CIRIA Special Publication 141. A joint CIRIA and Environment Agency pack that includes a training video, booklet and poster
- *Waste minimisation and recycling in construction: a review,* CIRIA Special Publication 122. A detailed report of waste minimisation and recycling in construction
- *Waste minimisation in construction – site guide,* CIRIA Special Publication 133. This outlines current good practice in site waste management and contains information on reducing wastage of raw materials, and reusing and recycling waste materials
- *The observational method in ground engineering: principles and applications,* CIRIA Report 185. This is specifically relevant to minimising waste in ground engineering and to optimising design to foresee problems on site
- *Managing materials and components on site,* CIRIA Special Publication 146. A CIRIA site guide that provides practical guidance for site managers, site engineers and supervisors on how to manage materials and components effectively.

Benefits and obligations

1.1 Introduction   **14**

1.2 The benefits of good practice   **15**

1.3 Environmental obligations   **16**

Introduction

Construction sites are often criticised for the damage they cause to the surrounding environment. This damage can take many forms, for example excessive noise or pollution from dust. Most people accept that everyone on site should follow environmental good practice. It is always more beneficial in the long term to follow good practice rather than bad practice.

Efforts are being made at all levels within the construction industry to implement environmental improvements:

- those in the boardroom are demonstrating their commitment by preparing environmental policies for their company
- clients are requesting evidence of environmental credentials from contractors before awarding contracts
- those on site are already implementing a variety of environmental initiatives.

However, action is still needed to improve the environmental performance of the construction industry, which, for example, causes more water pollution incidents than any other industry.

There are two major incentives for improving performance:

- enhanced environmental conditions resulting from good practice
- costly implications of the penalties for failing to meet environmental obligations.

These are explained in the following two sections.

Environmental benefits

Better environmental performance produces a number of benefits.

Reduced damage to our natural environment – Construction activities have the potential to cause damage to the surrounding air, land and water resources and to plant and animal life. The site team is responsible for minimising this damage.

Reduced demand for natural resources – The construction industry is a major user of resources that are mined or quarried from the natural environment. These resources are not renewable and their extraction causes damage to the environment. Reducing demand through better materials management, less wastage and greater use of reclaimed materials will conserve resources and reduce adverse environmental effects at source.

Reduced disturbance to our neighbours – Living near a construction site can be both disturbing and worrying. By complaining or by taking legal action, local residents may delay the project and increase costs. It makes sense to be a good neighbour both for their sake and yours.

Economic benefits

The economic benefits of environmental good practice are undeniably important. Implementation does not need to be costly; the truth is that sound environmental practice makes sound economic sense. This is clear in many elements of a contractor's business, resulting in:

Improved opportunities to tender – Clients in the UK and Europe are increasingly choosing contractors that can demonstrate sound environmental performance. A track record of prosecutions can damage a contractor's chances of being invited to tender.

Less money wasted on fines – Fines for environmental offences are increasing. This is bad enough, but the real cost of an environmental prosecution can be 20 times the cost of the fine levied when legal fees and management time are taken into account. It is even expensive to defend prosecutions successfully.

Less time and money spent repairing environmental damage – All spillages need to be cleaned up. Polluted rivers may need to be restocked with fish and damaged trees replaced. Cleaning up polluted groundwater is very expensive. In many cases clean-up can delay the project's progress; permission to resume construction work will not normally be given until a spillage has been cleared.

Less money lost through wasted resources – It is estimated that on some sites up to 10% of raw materials end up as waste. Money is wasted not only in buying the materials in the first place, but also through having to pay to dispose of them in a landfill. Landfill costs are rising due to landfill tax.

> *Preventing pollution is cheaper than curing it.*

Improved environmental performance will result in an improved environmental profile – This will help to establish good relationships with regulators like the Environmental Agencies and the local authority, thereby helping to ensure that projects run smoothly. It will also assist in developing staff morale and make it easier to recruit and retain good staff.

Benefits are felt at both a corporate and a project level. So what do they mean to the individual?

- The site manager can demonstrate improved margins.
- The site engineer's workload can be reduced by resolving conflicts.
- Construction personnel can enjoy the benefits of increased profit.

Environmental obligations

In addition to the incentive of gaining benefits from good environmental practice, there are a number of controls that demand that good practice is followed. These have both legislative and contractual origins and include:

Specification and contract conditions – These will have been drawn up to address any conditions imposed on the contract through the planning system and commitments made by the developer to the local communities. They may also include provisions made in an environmental assessment for the project. Failure to comply will be penalised through the contract.

National legislation – National legislation is in place to protect both the natural environment and the construction site's neighbours. Legislation such as the Environmental Protection Act 1990 and the Water Resources Act 1991 is policed primarily by the Environmental Agencies.

Other legislation is in place to protect specific features of the environment; under such legislation sites may be designated and protected by virtue of their ecological, archaeological, geological or geomorphological interest.

Conviction in the Crown Court (High Court in Scotland) can lead to unlimited fines and/or imprisonment for the person responsible.

Local control – There may be a number of requirements imposed by local authorities through the powers given to them by national legislation. These would include, for example, noise controls under the Control of Pollution Act 1974.

Guidance on relevant environmental legislation is provided in Chapters 3 and 4.

Corporate control – Many contractors have corporate environmental policies and environmental management systems that the sites are required to follow. In addition there may be a specific environmental plan for a particular project.

Most major construction companies have an environmental policy that requires the site to adopt controls to minimise environmental damage.

2 General site management issues

2.1 The management framework — 18

2.1.1 Setting the scene — 18
2.1.2 Dealing with regulatory agencies — 20
2.1.3 Management responsibilities — 21
2.1.4 Managing your contractors — 22
2.1.5 Raising awareness — 23
2.1.6 Liaising with clients and designers — 24
2.1.7 Environmental management systems — 25

2.2 Setting up and managing the site — 26

2.2.1 Management and site control — 26
2.2.2 The value of good public relations — 29
2.2.3 Site security — 31
2.2.4 Managing materials — 32
2.2.5 Traffic and access routes — 33
2.2.6 Site clearance following completion — 34

The management framework

This section outlines a framework for managing the environment on site. Steps 1 to 6 below should be followed for every site.

2.1.1 Setting the scene

Effective environmental management on site requires a team effort. This includes inputs from the main contractor and sub-contractors on site, the contractors' organisation off site and third-party organisations involved in the project (eg designers, clients and suppliers). To manage this teamwork effectively the **site manager** should follow the steps outlined below. So should the managers of any sub-contractor on site that has the potential to cause environmental effects – which applies to nearly all trades and operations.

STEP 1 *Establish the environmental obligations of the project*

STEP 2 *Identify the environmental hazards particular to the site*

- Initially, Steps 1 and 2 may be developed by reviewing the relevant documentation (Section 1.3). The site manager can add to this using the site team's experience, calling in specialists if necessary. It is useful to ask the regulatory agencies about their concerns for the site. Section 2.1.2 provides further guidance on this.

- It may be useful to liaise with clients and designers to establish how they can help to identify and overcome potential environmental difficulties (see Section 2.1.6).

- The distribution, review and monitoring of the information gathered during Steps 1 and 2 may be easier if it is collected together in a single reference document for the project.

STEP 3 *Establish an environmental management structure and plan*

- The site manager should define the environmental responsibilities of all personnel within the site management structure. This should include those who are involved in implementing and monitoring initiatives. It should also define lines of communication between staff and third parties, and the responsibility for producing the site environmental plan (see box opposite). Ideally, the latter would have a section addressing each of the environmental issues presented in Chapter 3. The level of detail in the plan will depend on the size and complexity of the project. It need not be a separate document but could be integrated with existing plans such as those for health and safety or quality.

Section 2.1.3 provides examples of management responsibilities that are defined in some contractors' organisations, and 2.1.4 discusses how to manage sub-contractors. Responsibilities for implementing emergency response procedures are presented in Chapter 3 for each environmental issue.

> The better that responsibilities are defined and understood the more likely they are to be taken seriously at all levels.

STEP 4 *Establish where environmental sensitivities are likely to require special construction procedures*

- The need for special construction procedures will vary from site to site and is likely to include elements such as the training of personnel and method statements defining the use of specialised plant or the provision of specially designed temporary works. Issues will generally have been considered in the pre-construction planning stage, but they may have been missed. Although this handbook does not cover the pre-construction stages, it may provide enough information to enable site managers to determine when special measures would be necessary.

Site environmental plan

When the procedures and outcome of Steps 1 to 4 are written down, this will constitute an environmental management plan for the project. These are already in use on many sites – they come in a number of forms and levels of detail. Their most important features are that they are accessible, regularly revised and in regular use.

Section 2.1.7 outlines how the site environmental plan can fit into the environmental management system of the company.

STEP 5 *Train your personnel*

- Environmental responsibilities need to be in place at all levels of an organisation since it only takes one act of ignorance or non-compliance to cause an accident. Appropriate training of personnel at all levels and clear definition of responsibilities will help minimise the potential for accidents to occur. Section 2.1.5 provides further suggestions on training.

STEP 6 *Monitor actions and effects*

- Monitoring is important to assess whether responsibilities are being fulfilled and to determine whether the environmental effects of the scheme are acceptable. Environmental monitoring may be carried out in parallel with that for health and safety and quality. Chapter 3 provides guidance on monitoring.

The management framework

2.1.2 Dealing with regulatory agencies

Whatever the size of the project it will be necessary to deal with regulatory agencies. These have a diverse range of specific responsibilities and powers to enforce legislation. Their responsibilities are described in the box below.

Regulators	Responsibility
Local authority	Noise, air quality, traffic, considerate contractors schemes
Environmental Agencies	Waste, effluent discharges, abstraction licences (England and Wales only), some nature conservation functions, ground contamination
English Nature	Designated ecological sites, geological and geomorphological sites, protected species
Countryside Council for Wales	
Scottish Natural Heritage	
County ecologist	
County archaeologist	Designated archaeological and heritage sites
Heritage bodies	
Health and Safety Executive	Health and safety
Sewerage undertaker	Effluent discharge to public sewer

It is always advisable to contact the regulatory agencies as early as possible to discuss the project with them. Regulatory agencies welcome an early approach and may be able to advise on environmental issues of local importance that should be included in the assessment of the project's environmental sensitivities (Step 4 of establishing the environmental framework for the site); this is particularly relevant on projects where an environmental statement has not been produced (ie the vast majority of projects).

It is important to develop a constructive dialogue with the officers of regulatory agencies that are monitoring the project. Explain what is happening on the project and why. This may simplify approval procedures and is likely to be helpful if things do go wrong.

If problems arise site personnel should always follow the company procedures specified in the site environmental plan when alerting the appropriate regulators. This handbook identifies when contact with particular regulators is appropriate.

- *Plan ahead. Try to avoid problems and give regulators advance warning of potential problems.*
- *Give regulators the time they need to process your enquiry.*
- *Always display the relevant emergency number for the regulatory agencies.*
- *Make sure that personnel know the correct procedures for reporting incidents – they should let the site manager know before contacting the regulatory agencies.*
- *Always notify the Environmental Agencies of any contaminating spillages.*

2.1.3 Management responsibilities

Ensuring environmental good practice is no different from any other task on site; in order to undertake it properly and comply with legislation, responsibilities need to be defined and understood by everyone. A sound approach is to ensure that environmental good practice starts at the top of the company in a similar way to good health and safety practice. Certain individuals have clearly defined roles (as outlined below), but everyone on site is responsible for ensuring that their actions constitute good practice. A typical working arrangement for environmental management on site is presented below.

Full-time site-based environmental posts are not usually necessary because responsibilities for environmental issues can be included within the responsibilities of existing staff. However, increasingly, medium to large sized contractors are creating a full-time head office post for an environment manager.

On most projects, regardless of the size, the **site manager** (or site agent) has principal responsibility for environmental management on that project. They may decide to define, monitor and control environmental practice themselves on site, or they may select a **delegated representative** to act on their behalf. The responsibility for environmental management will include auditing environmental practice on site, liaising with regulatory authorities and monitoring sub-contractors. Many companies have taken the opportunity of using their safety inspectors as environmental inspectors as well. The benefits are that a pool of expertise is developed and good practice can easily be shared around sites.

Site engineers and/or **site foremen** are usually in the best position to put the project environmental plan into practice. They need to understand the environmental obligations and the practical measures needed to comply with them. On large sites the plan can be subdivided, with a different engineer responsible for each section. For example, one person may be designated as the site waste manager, another may be in charge of noise control and so on. The engineers should then provide feedback to the site manager. In this way a consistent and thorough approach can be brought to the site as a whole.

The management framework

Site foremen and **supervisors** should ensure that environmental controls are implemented at the work face. They can best undertake this responsibility through working closely with the gangers. Together the foremen and gangers should review what training site personnel need and be instrumental in arranging for it to be provided.

On all sites, irrespective of their size, **all site personnel** must be charged with following good practice and encouraged to provide feedback and suggestions for improvements.

The responsibility for environmental good practice does not rest at the site level alone. In larger companies a main board member may be responsible for directing and reviewing corporate environmental protocols and responsibilities. In smaller companies this responsibility may be held by the **managing director.** It is good practice to appoint someone responsible for providing corporate advice on environmental legislation, good practice and the company's environmental policy. Whatever the company structure, it is vital that **intermediate management** translates decisions into action at site level.

The management responsibilities presented here for main contractors can easily be modified to apply to all sub-contractors, whatever their size.

> Successful environmental management relies on communication. It is crucial that everyone is aware of the key issues, has the relevant information to deal with them, understands their responsibilities, and provides feedback to those in charge.
> Site personnel must know whom they can contact for advice on managing environmental issues and whom they can ask for training. Feedback down the chain is important in maintaining motivation and raising awareness.

2.1.4 Managing your sub-contractors

The environmental awareness of sub-contractors varies considerably. Some may be particularly conscientious and cause no problems, while others may assume that all responsibility rests with the main contractor on site. The task of the site manager is to ensure that sub-contractors understand their obligations and meet them. As with any controls, environmental responsibility can be implemented with incentives or penalties. The following checklist presents suggestions for selecting, motivating and managing sub-contractors.

> You may be held responsible for your sub-contractor's offences if you do not exercise reasonable control over them.

Checklist

- *Ensure that sub-contractors are aware of the approach to environmental management on site, before commencing work.*

- *If the sub-contractors work frequently with the main contractor, then it is common for the main contractor to invite (or require) them to attend environmental training sessions.*

- *Ensure that they are aware of the environmental obligations of the project.*

- *When selecting sub-contractors, ask them to present proof of their past record in achieving good environmental practice. Stipulate that records of environmental prosecutions will be taken into account.*

- *Include controls in the sub-contract to encourage good environmental practice.*

2.1.5 Raising awareness

It is important to raise awareness of environmental issues so that people on site understand what environmental good practice is and know where to obtain information.

> Sound environmental management comes from knowing what to do and developing the right attitudes.

Training can be provided company-wide to spread good practice guidance generally, or at a site level to address issues relevant to a particular construction project. It can also be a mixture of both. For a successful training programme, it is vital to:

- select the right trainer
- tailor the level of detail of information to the audience
- select an appropriate mix of audience
- ensure that the material presented is relevant and focused
- follow up and reinforce training if necessary.

The management framework

> ### Key aspects of training for site managers and engineers may include:
>
> - relevant legislation
> - the reasons for following good practice
> - importance of good housekeeping
> - good practice in dealing with potential pollutants (eg oil refuelling/handling of paints and solvents)
> - emergency procedures
> - how to manage wastes
> - how to manage materials and components on site to reduce wastage
> - choice of plant
> - choice of working methods, and sources of advice
> - personal responsibility/liability.

It is useful to give site personnel brief environmental training as part of their induction to health and safety on site. However, in order for training to be effective, adequate time should be set aside to inform people about the issues that are relevant to their site and to their work. Toolbox talks can be a good way to do this. CIRIA, in conjunction with the Environment Agency, has produced a video, leaflet and poster set that may be useful for on-site training – *Building a Cleaner Future*, CIRIA Special Publication 141.

2.1.6 Liaising with clients and designers

> If you feel that the design of the temporary or permanent works compromises environmental good practice refer it back to the designer or client.

Although the designer should have included all environmental obligations in the contract documents, drawings or specification, it is worth checking whether any information has been inadvertently omitted. For example, often legislative acts are quoted without the accompanying obligations being identified.

A meeting between the designer and contractor on-site at the start of the contract may be useful to help the contractor understand where environmental issues may have affected the design.

Where the contract documents impose restrictive environmental conditions it is often worth discussing with the designers or specifiers the reasons for the conditions. It may be possible to explore alternative approaches that have the same (or even lesser) environmental impacts. For example, can bored piles be used instead of driven piles to reduce noise and vibration?

Where a project has undergone a full or partial environmental assessment there will be a considerable amount of information on the environmental issues surrounding the site. In these circumstances it is worth requesting that the client or designer gives a presentation on the environmental background to the project.

On some types of contract (eg DBFO), the client and designer may be within the contractor's own organisation.

2.1.7 Environmental management systems

It is important to have a mechanism to distribute instructions and information throughout the company, to check whether instructions are being followed, and to check that the objectives are being achieved. For environmental issues, an environmental management system (EMS) can be used as the mechanism.

An EMS is the mechanism by which a company sets, monitors and achieves environmental targets. This includes the definition of management responsibilities and reporting procedures. The site environmental plan (introduced in Section 2.1.1) can structure the management and reporting procedures on site to parallel those of an EMS. If an EMS is in place, the primary purpose of the site environmental plan is to focus on the specific environmental issues of the site.

There are two main international standards for environmental management systems in use in the UK – ISO 14001 and EMAS (European Community Eco-Management and Audit Scheme). In many ways these systems are very similar to the quality management system (BS 5750/ISO 9000).

Setting up and managing the site

Although the majority of environmental concerns on site arise as a result of the actual construction works, the general manner in which the site is managed can influence considerably the implementation and success of control measures.

This section outlines measures that the site manager should consider when establishing and running the site in order to provide incentives to staff, to reduce impacts and to reduce complaints from neighbours.

2.2.1 Management and site control

The vast majority of environmental accidents or causes of complaints stem from one or more of the following reasons:

- ignorance
- negligence
- carelessness
- vandalism.

It only takes one person to do the wrong thing for the individual or the company to end up in court.

A priority of the site manager is to ensure that site personnel understand that environmental issues must be taken seriously and that poor practice will not be tolerated. This can be achieved through a combination of setting a good example, rewarding good practice and punishing poor practice. Use the following checklist.

Checklist

- *Has an environmental plan been formulated and have ideas been developed for its implementation?*
- *Have environmental responsibilities been defined?*
- *Is everyone on site aware of their responsibilities and liabilities (including sub-contractors)?*
- *Are all environmental standards and obligations clearly defined?*
- *Have the standards been brought to the attention of all concerned?*
- *If training is necessary, has a training programme been established?*
- *Are environmental awareness posters/bulletins displayed?*
- *Are warning signs displayed prominently on site?*
- *If in place, is the company environmental policy displayed?*
- *If one exists, is the company environmental manual available?*

Considerate contractor scheme

A useful tool in demonstrating the site's environmental intentions is to work within a good practice framework, such as a considerate contractor scheme. These often involve adherence to a code of good practice, visits by external auditors, establishing good relations with neighbours, and incentive awards for tidy sites. Considerate contractor schemes provide an ideal framework within which to manage environmental issues but do not replace the need for an environmental management system (see Section 2.1.7). They are often administered by the local authority, although the Construction Industry Board has set up a nationwide scheme.

Case Study

Construction Industry Board's Considerate Constructors Scheme

The Considerate Constructors Scheme is a voluntary code of practice, driven by the industry, which seeks to:

- *Minimise any disturbance or negative impact (in terms of noise, dust and inconvenience) caused by the construction sites to the immediate neighbourhood*

- *Eradicate offensive behaviour and language from construction sites*

- *Recognise and reward the constructor's commitment to raise standards of site management, safety and environmental awareness beyond statutory duties.*

Good housekeeping

Good housekeeping is an important part of good environmental practice and it helps everyone to maintain a more efficient and safer site. The site should be tidy, secure and have clear access routes that are well signposted.

Checklist

- *Segregate waste as it is produced and remove waste frequently from site.*

- *Keep the site tidy and clean.*

- *Ensure that material and plant storage areas are properly managed.*

- *Keep hoardings tidy – repair them and repaint them when necessary, remove any flyposting.*

- *Frequently brush-clean the wheel washing facilities.*

Setting up and managing the site

General site appearance is important. When planning the site layout, contractors' offices and equipment should be sited to minimise visual intrusion. In rural areas, hedges, existing tree screens and the natural landform can be used to screen sites and compounds. In urban areas the construction works are usually screened with suitable hoarding.

Case Study

For a high-profile construction site in central London, in an area visited by many tourists, this impressive site hoarding was designed.

The hoarding was in character with the neighbourhood and provided a professional interface with the public.

Working hours

Site working hours can create considerable concern and annoyance among neighbours. On some projects, working hours for noisy operations are defined by the contract documents or by local authorities – perhaps through Section 60 or 61 notices (see Section 3.3). However, there may be opportunities for extending working hours in consultation with the local authority or the client. Although extensions to working hours may be crucial to the programme, their effect on neighbours should be considered carefully – try to restrict working to sociable hours. When extended working is needed, it is important to inform neighbours in advance of the reasons for the work and its duration.

Time activities within the allowable day carefully. For example, in the same way that it is advisable to schedule deliveries outside of rush hour, other intrusive activities can be scheduled at less sensitive times. To understand the constraints, which will vary from site to site, it is important to understand the daily patterns of the neighbours. Some points to consider include:

- in city-centre sites night-time noise may be more acceptable than day-time noise if there are no residential areas
- avoid noisy activities during school hours if possible
- local restaurants often appreciate less disturbance over lunchtime
- establish whether local businesses require quieter periods during the day
- determine whether weekend or night-time working is especially sensitive
- establish whether there are particularly sensitive areas such as hospitals near the site.

2.2.2 The value of good public relations

In the drive to complete the project with the minimum of disturbance to neighbours good public relations is vital. Experience has shown that members of the public tend to complain less if they know what is happening on site. Therefore, be prepared to explain the project and to answer their questions. Try not to change what the public has been told, as this can cause confusion and annoyance.

> Remember that you represent your client on site – your client's environmental reputation is at stake.

Public liaison is particularly important when operations that cause disturbance are being carried out for a significant period of time. Try to explain the efforts that are being made to limit the impacts of operations by phasing and other control measures.

Establishing good public relations is easier if the site personnel understand the project and its impact from the public's perspective. It is useful to walk outside the site boundary and view the site as a member of the public.

Case Study

On a particularly controversial road construction site, the site manager used contact with local schools as one of the site's initiatives to keep the community informed about the scheme. A detailed information pack was prepared for each of the local schools to use as a teaching aid.

Setting up and managing the site

✓ Checklist (some of these measures may be appropriate to larger sites only)

- *Visit occupants of particularly sensitive buildings and keep them informed of progress.*
- *Prepare a leaflet and distribute it to nearby residents or occupiers. Provide updates.*
- *Identify key local community representatives, such as parish councillors, and keep them informed.*
- *Write articles about the progress on site in the local press.*
- *Display a "Contact Board" at the site perimeter so that the public know who to contact if they have a complaint or a comment to make. Use this board to display information on the phasing and other relevant matters.*
- *Establish a complaint line and check that it works by calling it.*
- *Deal with any complaints that arise quickly and in accordance with a defined complaints procedure. Create a log of complaints.*
- *Consider providing a position so that passers-by can observe activities on site.*
- *Be able to identify your neighbours and understand their views.*
- *Join a considerate contractors scheme (see Section 2.2.1).*

Site personnel off site

Be aware that good relations can quickly be undone by the actions of the site personnel (including sub-contractors) off site. Causes of annoyance include:

- noise on arriving on site or leaving site
- overbearing numbers in local facilities such as pubs and shops during lunchtime and evenings
- vehicle parking.

Construction personnel can be very visible in the local community. Be alert to any antisocial behaviour or complaints and ensure that all staff know what is expected of them.

Lighting

Lighting can be an important deterrent to vandals and thieves on sites, but it can also annoy the local residents. Keep any site lighting at the minimum brightness necessary for adequate security and safety. Locate and direct the lighting so that it does not intrude on nearby properties. Remember that high levels of lighting result waste both energy and money. Consider using infra-red lighting for security.

2.2.3 Site security

Contractors can be held liable for environmental damage caused by vandals if they have not made reasonable attempts to guard against it. A contractor's liability increases if vandals have already struck once on a site.

Site security is an important component of good environmental management. Vandals often cause damage that harms the environment, by:

- opening taps on tanks containing fuel
- tipping out other liquids from drums and containers
- smashing/stealing raw materials
- playing on plant
- spraying graffiti or flyposting on site hoardings
- destroying works in progress.

The incidence of vandalism is higher on sites in urban areas, especially where they are close to schools or housing estates. Help to reduce vandalism by securing the site, and moving valuable items and those prone to theft from public view. Store these goods in a locked container or storage area. For information on protecting oils and chemicals stored on site from vandalism, see Chapter 4.

A contractor incurred a £4000 fine for polluting the Bulstake Stream at Oxford with gas oil following vandalism of an oil tank.

Suggested security measures

- Secure the site boundary using perimeter fencing and high-quality locks on gates. Various types of fencing are available and each has its own advantages and drawbacks. For example, solid barriers (eg hoardings) are more difficult to scale than chain-link fences and prevent casual surveillance by prospective thieves. However, they also provide cover to thieves and vandals once they are on site
- avoid stacking materials against the site boundary/fence as this can provide an opportunity for vandals and thieves to scale it

Setting up and managing the site

- within the site ensure that materials that are potentially hazardous to the environment are well secured. It is important to lock fuel outlets when not in use
- secure plant to prevent vandalism
- immobilise plant and equipment overnight
- install deterrents such as lights, warning notices, 24-hour security guards and alarm systems
- control the movement of people on and off site: use site passes or swipe cards
- position the site manager's office to give a good view of the site
- if the site is large or at high risk from trespassers, consider installing CCTV cameras
- inform local police about the site and seek their advice on security.

Key Guidance

- *Dealing with vandalism – a guide to the control of vandalism, CIRIA Special Publication 91, 1994*

- *The secure site – an impossible dream?, Chartered Institute of Building, Occasional Paper 44, 1991*

2.2.4 Managing materials

Improving the management of materials and components reduces material wastage and increases site efficiency. The environmental benefits of reducing wastage include minimising resource use and the amount of waste sent for disposal. Where site personnel follow established procedures for managing materials and components there will be fewer incidents of spillages and contamination arising from incorrect storage or handling, and less damage to materials and components; this means less wastage of raw materials.

Ordering and receiving deliveries

- Order the right quantity and quality of materials to arrive at the time when they are needed. This reduces the length of time materials have to be stored on site and therefore reduces the potential for damage or theft to occur
- when ordering, find out in what form the materials will be delivered, so that the appropriate unloading plant can be arranged
- after placing an order, check the arrangements for handling and storing materials as soon as they arrive on site
- always make sure that deliveries are received by a member of site personnel who is able to supervise the delivery, carry out a quality inspection and ensure that the materials are unloaded to the appropriate place.

Storage

There is usually a combination of central storage and workplace storage used on site; the balance between them depends on the site and the works in progress. It is important to manage storage areas well because they set an example for the site. Keep the following points in mind:

- ensure that the material suppliers' instructions on storage are being followed
- store materials that are valuable or attractive to thieves in a secure area
- store materials away from waste storage containers and from vehicle movements that could cause accidental damage
- secure lightweight materials to protect them from wind damage or loss
- take special care over the storage of materials that are potentially polluting. For information on how to store oils and chemicals on site, refer to Chapter 4.

Handling

There are many methods for moving materials around site. Options include cranes, trucks, fork lifts and even manual handling. Ensure that the suppliers' instructions on handling their materials are followed to minimise damage to materials and injury to site personnel.

Key Guidance

- *Managing materials and components on site, CIRIA Special Publication 146, 1998*

2.2.5 Traffic and access routes

It is important to manage site traffic, because it can cause delays to local traffic and create a safety hazard both on and off site. People living and working near the site are also often annoyed by the emissions, noise and visual intrusion of queuing vehicles.

Access routes

The use of public roads for site access may be restricted. Such restrictions may include weight and width controls, parking controls, steps to minimise pedestrian conflict and low-headroom access routes. Consult with the local police and local authority to address these issues and agree on a workable site access that does not compromise public safety. Plans may be required identifying each access point, the agreed route to the nearest main road, and the routes to be used by lorries to the road network. Wherever possible, arrange the access so that lorries enter and exit the site in a forward direction.

Setting up and managing the site

Managing site traffic

Plan the timing of deliveries to avoid vehicles waiting. Where several deliveries are likely to take place over a short period, designate queuing areas. In summer avoid queuing outside buildings without air conditioning as they often have open windows. In urban areas it may be best to allocate a waiting area some distance from site and then call in deliveries when access to the site is clear.

Site personnel car traffic often annoys the public. Arrange designated parking areas, ensure that staff do not park in unsuitable areas and ensure that restrictions are complied with. Consider implementing a park-and-ride or car-share scheme.

Sometimes construction sites are blamed for disturbance caused by vehicles that are not even associated with that site! To avoid this, it may be helpful if site vehicles display some visible identifying marks. While this may not be appropriate for individual deliveries it can be done for the main contractors' vehicles and for regular delivery vehicles.

Checklist

- *When ordering deliveries, ensure that all drivers are aware of traffic restrictions at and around the site.*
- *Arrange deliveries to site so that vehicles can go straight in without having to queue outside the site boundary.*
- *Instruct drivers to switch off engines when vehicles are waiting.*
- *Consider the use of in-cab communication systems to maintain control of lorry movements.*
- *Load and unload vehicles off the highway wherever possible.*
- *Plan parking for site personnel's vehicles.*
- *Consider getting regular site vehicles to display site identification.*

2.2.6 Site clearance following completion

This is a phase of the works that usually receives little attention, but it can cause most problems. In clearing the site it is vital that wastes are managed in accordance with the regulations (see Section 3.2). Normally, on completion of the works the contractor is required to clear away and remove from the site all plant, surplus materials, rubbish and temporary works. The whole of the site and works should be left clean – until then the project is unfinished.

3 Environmental issues

3.1 Water   36

3.2 Waste   48

3.3 Noise and vibration   57

3.4 Dust, emissions and odours   72

3.5 Ground contamination   80

3.6 Wildlife and natural features   85

3.7 Archaeology   94

Water

Why water management is important

It is vital to manage water properly on site to protect our environment. If watercourses are polluted, or unacceptable wastes are disposed of to the sewer system, you or your company may end up in court. Industries using water from a river downstream of site may be affected by reduced water quality and sue if this causes damage.

The site does not need to be next to a river to cause a problem. Any pollutants getting into a surface water drain or groundwater can end up in a river even if it is miles away. These pollutants can be tracked to their source.

> Spillages can be easily noticed. A gallon of oil can completely cover a lake the size of two football pitches.

Environment Agency research data shows the construction industry in England and Wales to have been the the biggest single industrial source of water pollution every year from 1994 to 1997; the industry was responsible for more than 500 pollution incidents each year.

It does not take much spillage to cause pollution. For example, the normal limits set by the Environmental Agencies on suspended solids are typically 30–40 mg/l. This is about the equivalent of mixing half a tablespoon of soil in a bath.

High levels of silt can clog up a fish's gills and eventually kill it. It can also smother invertebrates and sensitive plant life, which are themselves a food source for fish. When deposited on the stream bed, silt may prevent fish spawning and suffocate eggs. Levels as low as 180 mg/l can damage salmon and trout, and juvenile fish can be adversely affected by only 15 mg/l.

> In 1996 a contractor was found liable for causing oil pollution of the Grand Union Canal, after a tap was removed from an oil storage tank. In its defence, the company blamed vandals. Magistrates agreed with the Environment Agency that the site had poor security and the company was fined £7000 and had to pay costs of more than £11 000.

Other pollutants damage the water environment in other ways, for example, by changing the chemical balance (cement or concrete washwater is highly alkaline) or by removing dissolved oxygen (eg detergents). Contaminants that dissolve quickly are very difficult to control and treat. They are easily transported in watercourses and, if toxic, the effects are likely to be widespread. Notify and seek advice from the Environmental Agencies if spillages occur.

Managing water on site

A structured approach is required to manage water successfully to avoid causing pollution and to minimise cost and effort for all concerned. The following steps should be adopted.

STEP 1 *Evaluate the potential challenges and risks for the project*

Water pollution problems arise as a result both of activity and inactivity on site. Key causes of problems include:

- silty water and its incorrect disposal (see page 40)
- spillages of pollutants due to bad storage and handling of materials (see Section 4.18), or the inadvertent disposal to surface water drains rather than sewer (see page 41)
- washout from concreting operations (see Sections 4.4 and 4.5)
- works in, above or near watercourses (see Section 4.20)
- working with groundwater (see Section 4.21)
- water in excavations (see Section 4.9)
- works in contaminated ground such as dewatering, ground improvements, excavations and pumping (see Section 4.21).

STEP 2 *Identify appropriate control and management methods for each potential issue identified in Step 1*

This section of the handbook (3.1) outlines some approaches that can be adopted.

STEP 3 *Ensure compliance and monitor implementation*

Water pollution incidents usually arise as a result of ignorance, negligence or vandalism, so guard against each of these. Therefore:

- predict potential pollution incidents by undertaking risk assessments
- provide training to eliminate ignorance
- combat negligence by supervising site personnel
- secure sites against vandalism (see Section 2.2.3).

STEP 4 *Adopt an emergency response plan*

Even with the best controls in place pollution incidents may still occur. Therefore ensure that the site has an emergency response plan and that all relevant site staff know how to put them into practice. See page 44 for guidance.

Water

General good practice

Know your site

Before starting works consider carrying out surveys to establish the existing water quality and water levels around site. This data may be useful if an incident occurs. Establish the location of local streams and mark the position of surface water drains and foul sewers on site. Use colour coding to distinguish them – blue for surface water and red for foul water. Drains can lead to controlled waters, so be careful what you discharge to them.

Look out for underground pipes and tanks. Avoid disturbing pipes, particularly those containing gas (yellow) or foul water (brown/black) or diesel.

Carry out regular inspections of all discharges, drainage systems, collection ditches, lagoons, interceptors and watercourses to check that these are in good order.

> Damage to drains and sewers by the use of heavy plant or through digging can cause blockages that may lead to pollution incidents. Protect services before tracking plant over them. There have even been pollution incidents resulting from sewers that are blocked with timber and safety hats!

Abstracting water

In some instances it may be necessary to obtain a licence to abstract water for use on site for applications such as concrete batching and dust suppression, for example. Contact the Environmental Agencies for information.

Disposing of water from site

This applies to a wide range of types of discharge, both polluted and unpolluted. Construction site runoff and all waste waters arising must be disposed of in accordance with the requirements of the relevant regulatory authorities, for example:

- consent is required from the local sewerage undertaker to discharge effluent to public foul sewers
- consent is required from the Environmental Agencies to discharge direct to a watercourse. The consent will establish allowable concentrations of pollutants and flow rates. It may even prescribe peak flow rates for unpolluted discharges.

In some circumstances consents may take time to obtain, so plan ahead to avoid delays. Even with a consent, allowable pollution limits will be low. Controlled waters include: rivers, streams, ditches, ponds, groundwater. Pollution includes: silty water, oils, chemicals, litter, mud.

Before discharging water, therefore, check that you have permission to do so and that the discharge complies with any conditions attached to that permission. This may require monitoring of discharges (see page 46) to check that the discharge is within the consented conditions. For information on liaising with regulatory authorities, see Section 2.1.2.

Prevent anything that has the potential to pollute, including muddy water, from entering the surface water drains.

For information on the disposal options for silty water see page 40. As a last resort waste water may be pumped into a tanker and taken off site for disposal; however, this may be costly. If small quantities of waste water are generated, pump it into clearly labelled bowsers for removal off site at a later stage.

Avoiding spillages

There are many precautions that can be taken to avoid spillages. These are outlined in Chapter 4, and include the use of bunds around oil storage tanks (see 4.18) and the use of drip trays around mobile plant (see 4.19). Advance planning can avoid the need for emergency response

if things do go wrong. For example, sandbags, or even sand, can be used as a barrier to protect sensitive areas, or block off drains, during refuelling. They are also effective for controlling and mopping up spillages. Any sand or soil that becomes contaminated must be disposed of properly.

Sand placed around a drain to prevent water reaching a nearby watercourse.

Managing effluent from vehicle washing

Where possible, wash vehicles in a bunded area and recycle the cleaning water or discharge it to the foul sewer (with the consent of the sewerage undertaker). Another option is collection in a sealed tank for removal from site by a licensed waste disposal contractor. In exceptional circumstances, discharge to surface water may be permitted with the consent of the Environmental Agencies. However, this would require treatment (for example, silt settlement and oil separation) and it may be difficult to control quality.

> Do not wash tools and equipment in any watercourse.

Managing solid wastes

Prevent litter from blowing into watercourses by storing wastes properly. See Section 3.2

Managing surface water runoff

Surface water running across or ponding on a site may cause water management or pollution problems. For example more than 25 tonnes of sediment can be eroded per hectare of site in a year. The solution is to control surface water so that it does not run into excavations, disturbed ground or haul roads. See Sections 4.8 and 4.9 for guidance. Ensure that the water collection system is adequate to allow for the controlled release of storm flows. Protect watercourses from silty runoff from disturbed ground (from haul roads or after topsoil stripping) and soil stockpiles.

In dry weather large quantities of mud and oils can build up on areas of hardstanding. If these are not cleaned frequently, a sudden shower can wash them into watercourses, giving very high pollutant loads. Therefore keep hardstanding and surface roads swept clean.

> A silt fence can be used to prevent silt from entering watercourses. It comprises a vertical fence about 300–450 mm high made of two layers of woven geotextile that are welded together. Posts placed between the two layers support the fence. The fence should be dug in slightly, both to prevent runoff escaping underneath it and to give it extra strength.
>
> Silt fences can be placed for example at the toe of stockpiles (about 5 m away from the base) or at the top of banks of watercourses. Collected silt should be cleaned away regularly. They work best if slopes draining to them are less than about 30 m long. The ends should return up the slope to stop flows around the ends.
>
> Do not rely on them to remove sediment from concentrated flows such as pipe discharges and in watercourses.

Discharging silty water

Considering the options

To establish the best approach for your site, review the following prioritised options; those at the top of the list are least expensive and cause least risk of accidental pollution.

1. Pump to grassland or other excavation/soakaway – these areas should preferably be well away from excavations to avoid recirculation through the ground. The silty water should contain no chemical pollutants. (See [A] below.)
2. Pump to sewer (see guidance earlier in this section).

3. Pump to settlement tank (see [B] below).
4. Pass through a filtration system (see [C] below).
5. Use flocculants in conjunction with settling tank (see [D] below).
6. Pump into a tanker and dispose of off site.

The preferred option will depend on a number of factors, including:

- the quantities of water involved
- whether areas are available for storage and treatment
- the level of any charges to be levied by the sewerage authority
- the degree of contamination of the water
- the characteristics of the sediment.

Whichever system is applied, controls should always be in place to stop things going wrong.

[A]   Pumping to grasslands/fields

Obtain the approval of the Environmental Agencies and the landowner before doing anything. This option is really only suitable for water that is unpolluted aside from its silt content. It may also be necessary to allow water temperatures to rise by storing in a lagoon before discharge onto crops. Typical infiltration rates are given below for various soils.

Typical infiltration rates for various soils

Soil texture	Sand	Sandy loam	Loam	Clay loam	Silty clay	Clay
Infiltration rate (mm/h)						
Typical mm/h	50	25	13	8	3	5
(range)	(25–250)	(13–75)	(8–20)	(3–15)	(0.3–5)	(1–10)
Typical l/min/ha*	8000	4000	2000	1300	500	800

* Assuming the water to be spread evenly over the surface. A typical agricultural sprayer discharges at about 1000 l/min.

At one site it had been agreed that silty water would be pumped onto adjacent land and not direct to the river. A pollution control officer from one of the Environmental Agencies had visited the site and seen water being pumped direct to the river, but had not objected because the water was clean. This led to a misunderstanding that the method of disposal direct to the river was approved by the Environmental Agencies when clearly it was not. A subsequent prosecution was brought when silty water was discharged to the river.

Water

[B] Settlement tanks/lagoons

A settlement lagoon (or tank) works by retaining water in an undisturbed state long enough for suspended solids to settle out. The clean water then either flows out at the discharge point or is pumped out. It is usually necessary to retain the water in the settlement tank for several hours to reduce the level of suspended solids in the water. A rule of thumb indicates that a retention time of 2–3 hours at maximum flow rate is adequate. However, finer particulate matter (eg clays) may require a longer retention time and possibly larger lagoons. An idea of the required retention time can be obtained by leaving a sample of the effluent to stand in a milk bottle. Compare this sample to a prepared sample of a concentration equal to your consented discharge. As a general rule, when the effluent sample looks as clear as the sample representing the consented discharge, it may be discharged.

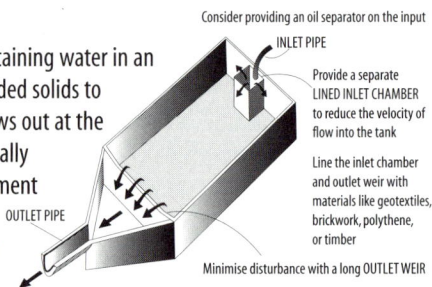

Consider providing an oil separator on the input

INLET PIPE

Provide a separate LINED INLET CHAMBER to reduce the velocity of flow into the tank

Line the inlet chamber and outlet weir with materials like geotextiles, brickwork, polythene, or timber

OUTLET PIPE

Minimise disturbance with a long OUTLET WEIR

✓ Checklist – settlement tank/lagoon

Design

- *The size is determined by the settling time required and the rate of introduction of water*
- *Keep waters as still as possible by incorporating the features shown in the diagram*
- *Examples of typical dimension of a settling lagoon for a three-hour settling time are given below*
- *A settlement lagoon that is long, narrow and shallow helps to ensure maximum retention time of all water in the lagoon. More efficient settlement may be achieved with two or three lagoons in series rather than one large lagoon, since this system may reduce disturbance of the water.*

Operation

- *It may be feasible to pump clean water from the surface of settling lagoons into rivers, but remember that water at lower levels will be more silty*
- *Clean the entry chamber periodically*
- *Carry out periodic monitoring of the outflow quality.*

Typical dimension of a settling tank/lagoon for a three-hour settling time

Pump diameter	Discharge rate	Length	Width
6 in pump	3000 l/min	60 m	20 m
	6000 l/min	80 m	27 m
4 in pump	1000 l/min	30 m	10 m
	2500 l/min	50 m	17 m

(Assuming 1 m-deep ponds, where the length = three times the width)

Consider installing a spare settlement tank on site, so that it can be used as an emergency overflow or when the main tank is being cleaned.

Surface water drainage from a site overflowed from three settlement tanks into a settlement lagoon and then to a consented discharge point in the River Thames (Dagenham Breach). There was a discharge of "very chalky water" from the settlement tanks. The contractor pleaded guilty to "causing polluting matter to enter controlled water". A fine of £1000 was imposed, although it could have been up to £20 000.

[C] Filtration

Discharges with fairly coarse particles may be treated easily and cheaply by passing them through steel tanks or even skips filled with a suitable filter, such as fine single-size aggregates (5–10 mm), geotextiles or straw bales as filters. If this solution is used there must be careful control of the effluent quality and a mechanism to close down the flow.

Water from a settling lagoon was allowed to drain to a culvert having first been filtered through a geotextile. This arrangement had been discussed with and agreed with the Environmental Agencies. One day, there was an increase in the flow from the lagoon and the geotextile was disturbed by persons unknown, allowing silty water to enter the river. In court, the contractor pleaded guilty to causing trade effluent to be discharged into the river.

[D] Flocculation

If no alternative treatment is available, due to the material characteristics, lack of space or lack of suitable discharge locations, then the use of settling tanks together with flocculant (or coagulant) can be an effective solution. This approach needs specialist advice on the design of the system, the size and dosing rates etc, and careful supervision to avoid failures of the system (as demonstrated by the example below). Flocculants commonly used include aluminium sulphate and ferric sulphate.

Remember that flocculants themselves can be highly polluting and toxic to river life.

Beware: fines can be substantial. A water company was fined a total of £175 000 in August 1996 for polluting the River Elm in Mid Wales. It pleaded guilty to two offences: the first allowed the entry of a coagulant (ferric sulphate) into the river, and in the second a noxious substance was allowed to flow into water that contained fish.

Water

Emergency response

Follow the emergency plan for the site. In the event of silting, erosion or pollution of a river, stream, ditch, other watercourse or water in underground strata, the site manager should call the emergency pollution line of the Environmental Agencies immediately on 0800 807060. Following clean-up, the incident must be reported to the company's environmental representative so that it can learn from what happened.

An effective emergency response system relies on the following elements:

- an emergency response plan
- contact numbers
- definition of responsibilities
- training in implementation.

Site managers should ensure the necessary information and equipment is at hand and updated regularly. This will require **advance planning!**

Emergency response plan

Ensure that all appropriate staff are aware of the company's emergency procedure and know how to use it (see the example given below).

EXAMPLE OF A TYPICAL EMERGENCY PLAN

1. In case of spillage of oils and chemicals report immediately to manager/supervisor, who should then report the incident to the Environmental Agencies and/or sewerage undertaker. They will find out eventually, but if you report first, it will build a better relationship.

2. Try to identify the source of pollution and stop the flow immediately.
   Switch off sources of ignition.

3. Avoid the spillage spreading   Check the site drainage plan – where will the spillage go?
   Stop the flow if possible
   Dam the flow with earth/sand/polythene
   Divert from drains/watercourses where possible

4. Get a spill kit
   Use absorbent materials if appropriate
   Place a boom across watercourses as precaution

5. Do not wash spillage into the drainage system – it only makes things worse. Never use detergents. Use sand or absorbent pads to mop it up.

6. If the spill has already entered the drains, block off the entrance to the drains.

7. Shovel contaminated sand/earth/granules into sacks or skips according to size. These must be disposed of appropriately. Oil pools may be removed by sludge-gulper first.

Responsibilities

Define the responsibility for the following with site personnel:

- reporting to the site manager,
- reporting to the Environmental Agencies and other regulatory authorities
- taking charge at the scene
- recording events as an incident record
- regularly checking that the contents of the spillage kits are complete.

Ensure that the contact details for the following groups are easily available:

- list of site personnel and sub-contractor offices
- your company's environmental representative
- fire-brigade/police 999
- Environmental Agencies – 0800 807060
- local authority environmental health department
- sewerage undertaker
- equipment suppliers (eg pump hire and waste disposal sub-contractors for skip hire)
- liquid waste disposal contractors.

Equipment

- Emergency spill kits are ideal for dealing with spillages, usually consisting of equipment to contain and absorb spills on land and water. Obtain them from a reputable supplier
- the contents will depend on the project, but they may include: oil-absorbent granules, "pigs" or "sausages", floating booms, absorbent mats, polythene sheeting, polythene sacks
- store them in a marked bag or wheely bin in a well-signposted location. It is best to store them near where they may be needed. Ensure in advance that booms for rivers are long enough and have suitable anchorages
- assess the number and deployment of kits for quick access across site.

Buckets of sand, earth, straw bales and rags are good for cleaning up small spillages. There is also a wide range of proprietary equipment available from suppliers to deal with spillages. Special mats or cushions can be placed over drains to prevent pollution to water supplies or located at the source of leaks or under pipe joints.

Training

Make sure that site personnel know who to contact in the event of a spillage, what to do and where to get equipment from. Manufacturers of spillage kits usually provide training in their use.

Water

How and what to monitor

The chemical analysis of water samples for trace quantities of contaminants or for particular pollutants can be a specialised task. However, much may be learned from the visual appearance of a water sample and this may be supplemented with simple on-site test kits to gain an appreciation of the polluting potential of a sample. The following can give an insight into the potential problems associated with a water sample:

- colour
- odour
- suspended solids
- presence of oil film.

The use of pH papers and test strips can also indicate the presence of specific components on site. For instance, high pH rinse waters from cement batching plants or grouting works can be toxic to aquatic systems and may cause a severe pollution incident. This can be easily identified through the use of pH papers which show a colour change in response to differing pH.

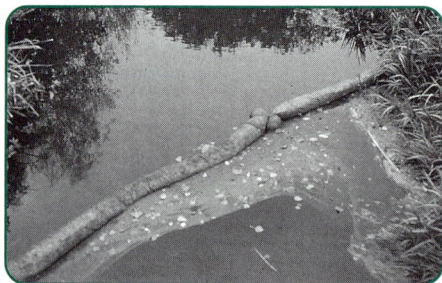

It makes sense to place booms permanently across very sensitive watercourses if spills are possible. Discuss with the Environmental Agencies where this may apply.

Note how booms are joined.

Key Guidance

The following Pollution Prevention Guidance Notes are available, free of charge, from the Environmental Agencies:

- *General Guide to the Prevention of Pollution of Controlled Waters*

- *The Use and Design of Oil Separators in Surface Water Drainage Systems*

- *Disposal of Sewage Where No Mains Drainage is Available*

- *Works In, Near or Liable to Affect Watercourses*

- *Working at Demolition and Construction Sites*

- *Safe Storage and Disposal of Used Oils*

- *Above Ground Oil Storage Tanks.*

Legislation

Water pollution

It is an offence to cause or knowingly permit any poisonous, noxious or polluting matter or any solid waste matter (which includes silt, cement, concrete, oil, petroleum spirit, sewage or other polluting matter) to enter any controlled waters unless a discharge is authorised. Road drains and surface water gullies generally discharge into controlled waters and should be treated as such.

> The maximum penalty on summary conviction is:
> - imprisonment for up to six months or a fine not exceeding £20 000 or both
> - on conviction on indictment, imprisonment for a term of up to five years or an unlimited fine or both.

Discharges to sewers

An occupier of trade premises (<u>which includes a construction site</u>) is committing an offence if trade effluent is discharged into a sewer without the sewerage undertaker's consent. The sewerage undertaker will usually be the company that also is responsible for the supply of water. As a condition of granting consent, the sewerage undertaker will levy a charge on the occupier of the premises based on the quantity and quality of the effluent concerned.

> In England and Wales the penalty on summary conviction is:
> - a fine not exceeding £5000
> - the penalty on conviction on indictment is an unlimited fine.

Environment Act 1995

The Environment Act 1995 further strengthens the provisions available to the Environmental Agencies to prevent pollution incidents, undertake anti-pollution works or serve a works notice on any person in situations where polluting matter has entered, or is likely to enter any controlled water. The works notice can require the person to:

- remove or dispose of the polluting matter
- remedy the pollution and restore the affected areas of water.

The Control of Pollution Act (1974 and amendments) in Scotland has the same provisions.

Other legislation

Other legislation to be aware of includes:

- The Salmon & Freshwater Fisheries Act 1975
- Land Drainage Act 1991
- Environmental Protection Act 1990.

Waste

What is waste?

Problems often arise with waste because there is confusion as to what waste actually is. Legally, waste is defined as "any substance or object which the holder discards, intends to discard, or is required to discard". Be careful – material that you do not regard as waste may fall under this definition, for example excavation spoil. If in doubt seek advice from the Environmental Agencies. Be aware that there are many different types of waste (see the box below) and these need to be treated differently.

Definitions of different types of waste

Inactive waste covers materials that do not undergo significant physical, chemical or biological reactions or cause environmental pollution when deposited at a landfill under normal conditions. These include uncontaminated soils and rocks, ceramics, concrete, masonry and brick rubble, minerals.

Active wastes are those that are not inactive wastes (see above). They include acids, pesticides, wood preservative, oily sludges, batteries, waste oils, asbestos, timber, plastics, alkaline solutions and bitumen. Some active wastes may also be special wastes. Active waste is subject to a higher rate of landfill tax than inactive waste.

Special wastes are those that are deemed to be dangerous to life; they may be corrosive, reactive, explosive, oxidising, carcinogenic or flammable. Some of the more common special wastes include acids, alkaline solutions, oily sludges, waste oils and wood preservatives. The criteria to be used to determine whether a waste is special waste is specified in the Special Waste Regulations 1996 and the technical guidance given in DETR Circular 9/96.

Why is waste important?

Compliance with legislation

Sites must comply with legislation concerning the storage, handling, transport and disposal of waste. Fines for non-compliance are increasing.

A contractor in Cambridgeshire was fined £2250 for illegally depositing waste material consisting of soil and other rubble excavated from a petrol filling station onto land in the area.

Damage to the environment

The UK construction industry generated over 70 million tonnes of construction and demolition waste (including excavation spoil) in 1991. This is about four times the rate of household waste per person. There are environmental impacts of dealing with these wastes, for example noise and traffic emissions, that would not arise if the waste was not produced.

Impact on project programme and budget

Handling waste badly will cost time, money and effort. For example, it takes time to rehandle waste that was not managed properly initially.

It costs money to buy the materials in the first place, to store the waste on site, to transport it from site, and to dispose of it. Landfill is costly.

People often think that waste costs are only the costs of its disposal. In fact, the true cost of waste is:

- the purchase price of materials that are being wasted
- the cost of storage, transport and disposal of waste
- the cost of the time spent managing and handling the waste
- the loss of income from not salvaging waste materials.

The main problems

The main problems are:

- wasting excess quantities of raw materials
- storing and handling waste badly
- disposing of too much waste to landfill
- not following the rules on waste transport and disposal.

Waste management should start with waste minimisation

Waste

Waste management on site

To manage wastes effectively, focus on:

- the amount of materials that are wasted
- the way in which wastes are handled and stored
- the amount of wastes that can be reclaimed
- the method of disposal of wastes.

This section provides guidance on how to address each of these issues. The site manager should allocate responsibility for these issues on site to <u>nominated individuals</u>. On larger sites it may be most appropriate to designate one person as the site waste manager.

To manage wastes effectively and efficiently it is important to allocate sufficient space and resources in advance. In order to plan waste management, it is helpful to know what type and quantities of wastes are generated on site. This information may be obtained either by monitoring wastes on site (see below) or by drawing on previous experience – bearing in mind that improvements to waste management practices will probably reduce the volume requiring disposal.

Monitoring wastes on site

The site manager (or a representative) should carry out waste audits at regular intervals to look at:

- the quantities of raw material wastage
- the quantities of waste of each type generated
- the way in which wastes are being handled and stored
- the costs of disposal for different types of wastes.

Carrying out audits will help to show how well waste management initiatives are working on site.

Wastage of raw materials

It is commonly accepted that an extra 5–10% of materials should be ordered to allow for site wastage through damage, spillage, over-supply and vandalism. These figures could be reduced. The box below can be used as a focus for looking at how materials are ordered, delivered, stored and handled on site to investigate how wastage can be reduced. Refer to CIRIA Special Publication 146, *Managing materials and components on site* for practical information on how to reduce wastage on site (see Key Guidance).

Reducing wastage of materials	
Ordering *Avoid:* ● overordering ● ordering standard lengths rather than the lengths required ● ordering for delivery at the wrong time	**Delivery** *Avoid:* ● damage during unloading ● delivery to inappropriate areas of site ● accepting incorrect deliveries, specification or quantity
Storage *Avoid:* ● exceeding shelf lives ● damage or contamination from incorrect storage ● loss, theft and vandalism	**Handling** *Avoid:* ● damage or spillage through incorrect or repetitive handling ● delivering the wrong materials to the workplace

Storage and handling of wastes

Aim to segregate different types of waste. Segregation has many benefits:

● it is easier to see what types of waste are being produced and where efforts to reduce waste need to be targeted

● it can reduce landfill tax payments, because the contamination of inactive wastes by active wastes is reduced

● it maximises the potential for reusing and recycling materials (see below).

Look at the checklist below for the key issues in segregating wastes. Consider wastes other than construction wastes: store canteen waste in covered containers and keep office waste separate from other types of waste. See the box on page 53 for information on dealing with special wastes.

Checklist – storing wastes properly on site

● *Segregate different types of waste as they are generated.*

● *Mark waste containers clearly with their intended contents. Consider using colour coding.*

● *Use containers suitable for their contents. Check that containers are not corroded or worn out.*

● *Minimise the risk of accidental spillages or leaks. Provide covers and bunds to prevent evaporation and spillage of wastes.*

● *Ensure that wastes cannot blow away.*

Waste

Case Study

Some contractors have developed a system for segregating waste on site using labelled bins provided by their waste management sub-contractor. The waste management contractor can offer reduced rates for disposal of segregated wastes because much of it can be recycled and the remainder attracts less landfill tax. This arrangement has been reinforced by a regional contract between the contractor and the waste management contractor for all their sites, which makes the investment in dedicated waste storage containers worthwhile.

Landfill tax

Waste disposed of to landfill is subject to landfill tax. This includes materials like waste soil and waste clay that are often sold to the landfill operator for use as daily cover. There are two bands of landfill tax – active waste is subject to a higher rate than inactive waste (see definitions on page 48). If these are mixed together the higher rate of tax will be charged on the whole load, so segregating waste saves money.

An incidental amount of active waste is permitted in a mainly inactive load, for example a piece of wood in a skip, or small quantities of bituminous material or grass in a load of soil.

Examples of unacceptable amounts of active wastes in mixed loads are:

- rubble from the construction of a house containing materials such as paint tins, unused tar and leftover plaster
- a large piece of wood in a skip or lorry, such as a roof beam.

Exemptions from landfill tax can be obtained for particular wastes, such as dredgings from inland waterways. Apply for exemptions before sending the waste to landfill.

Managing special wastes

Identify special wastes that are generated on site. Ensure that everyone who might be exposed to special wastes is aware of their potential effects and what to do with them.

Storage

- Store special wastes safely in containers and label them correctly
- avoid mixing special and non-special wastes to save creating large quantities of special wastes that will raise disposal costs
- avoid mixing different types of special wastes – unless it is known that mixing is harmless and will have no adverse effects on the subsequent management of the waste.

Transport and disposal

The Environmental Agencies must be informed before any special waste is removed (this is called pre-notification), using a special waste consignment note.
The consignment note must accompany transport of special waste and replaces the need for a waste transfer note.

- identify an appropriate disposal route
- store the documentation for the transport of special wastes safely.

Reducing the amount of waste going to landfill – look at waste as a resource

Minimise disposal costs by reusing and recycling wastes generated on site wherever possible. This will be easier to do if wastes are segregated as they are generated on site (see previous section). However, before stockpiling wastes on site for reuse or recycling, identify where the material is going to be used. If you do not have space on site to segregate wastes for reuse or recycling, consider off-site recycling by using a waste management sub-contractor that has the necessary facilities.

Case Study

On one inner-city office building site, roller bins were used to ease waste handling. Roller bins were filled up, marshalled in an area near the lift and then lowered to the ground level for emptying when the truck arrived daily. This was an ideal solution for this confined site because there was no room at ground level to store a skip permanently.

Waste

Examples of materials arising as wastes on site that may be reused or recycled are given below. Wastes arising off site may also be used on site. To find sources of reused and recycled materials, talk to local demolition contractors and local authority recycling officers.

Be aware that there may be restrictions on the reuse and recycling of certain materials in some applications; refer to the Environmental Agencies and the Key Guidance section for information.

Type of waste	Can it be reused or recycled?
Concrete	Recycle for use as aggregate in new concrete Recycle for use as unbound aggregate in roads or fill
Blacktop	Recycle for use in bound layer of road Recycle for use as bulkfill
Excavation spoil	Recycle for use as fill Reuse for landscaping
Topsoil	Reuse for landscaping
Timber	Reuse eg for shuttering/hoardings Recycle for chipboard
Metals	Reuse Recycle
Architectural features	Reuse
Clay, concrete pipes, tiles, blocks and bricks	Reuse Recycle for use as fill
Packaging and plastics	Recycle – ask the supplier

BRE (the Building Research Establishment) in Garston has set up a materials informations exchange to enable companies generating reusable or recyclable waste materials to get in contact with organisations able to use the materials.

Transporting and disposing of wastes

Unless the contract says otherwise, it is the contractor's responsibility to dispose of the waste arising from the project at a licensed and suitable disposal site, strictly in accordance with the Environmental Protection Act (1990) and the Environmental Protection (Duty of Care) Regulations (1991). If information is needed on the best disposal route for a particular waste (especially special waste), try contacting the supplier of the raw material (where relevant) or talk to the Environmental Agencies.

The person on site with responsibility for waste management must be able to describe both the waste kept on site and the waste transported off site. Waste leaving the site has to be accompanied by a waste transfer note (or special waste transfer note) that records the description of the waste, its current holder, the person collecting it and its destination.

Only sub-let the transport of waste to a registered waste carrier that is registered with one of the Environmental Agencies. If you have any doubts that your carrier is allowed to carry waste, ask to see its certificate of registration and check it with the Environmental Agencies.

Monitor the movement of waste off site to check that appropriate systems are being followed by using the checklist below.

Checklist – duty of care

- *Check that you have a copy of the carrier's registration document on site and that it is still valid. A photocopy of the carrier's registration document that has not been endorsed by the Environmental Agencies should not be accepted.*
- *The waste carrier must be authorised to carry that type of waste.*
- *The transfer notes should be completed in full and contain an accurate description of the waste.*
- *Keep copies of all transfer notes for waste sent off site.*

It may be useful to carry out a few spot checks on your waste carrier. For example, check that the waste does actually arrive at the agreed licensed destination. Check that the carrier does not return to site without sufficient time having elapsed for it to have reached the agreed licensed destination.

Key Guidance

- *Waste minimisation in construction – site guide, CIRIA Special Publication 133, 1997*
- *Landfill Tax Helpline: 0645 128484*
- *Local office of the Environmental Agencies*
- *Waste management – the Duty of Care, Code of Practice, HMSO*
- *Managing materials and components on site, CIRIA Special Publication 146, 1998*
- *Use of industrial by-products in road construction – water quality effects, CIRIA Report 167, 1998*
- *The reclaimed and recycled materials handbook, CIRIA publication C513, 1999*
- *Guidance on the disposal of dredged material to land, CIRIA Report 157, 1996*

Waste

Legislation

Environmental Protection Act 1990

Part II provides the main controls for waste management, including the introduction of the duty of care on all those who produce or keep waste to require them to prevent escape of waste, ensure proper disposal and ensure transfer to a registered carrier with a proper transfer note. Refer to the Environmental Protection (Duty of Care) Regulations (1991) for more information.

Special Waste Regulations, 1996

These regulations apply if special wastes are produced on site. (For a definition of special waste see the box above.) If a waste was defined as "special" under the 1980 Control of Pollution (Special Waste) Regulations, then it will be defined as special waste under the 1996 Regulations. Additional wastes, however, are also defined as special wastes, so refer to the guidance contained in Circular 6/96 from DETR. If the waste to be classified is still in its original container, look at its label – an orange and black symbol indicates that the material is special waste.

Carriage of Dangerous Goods by Road Regulations 1996 and the Carriage of Dangerous Goods by Rail Regulations 1996

These regulations require that special wastes are considered dangerous and include requirements that:

- suitable vehicles are used for transport
- emergency instructions in writing are available during transport
- appropriate care is taken during loading, storage and unloading.

Other relevant legislation

- Waste Management Licensing Regulations, 1994
- Producer Responsibility Obligations (Packaging Waste) Regulations, 1997.

What is noise?

dB(A) scale

Noise is often explained as being a sound that is unwanted by the listener.
Sound is a wave motion carried by air particles between the source and the receiver, usually the ear. It may consist of a high-pitched or low-pitched whine or it may have no special distinguishing features. Sound, pressure and noise are measured in units of decibel (dB) using a logarithmic scale. This means that if you increase a sound by 10 dB it is perceived as a doubling in loudness.

The diagram shows typical noise levels of everyday situations and the effects of noise levels.

As a rule of thumb, if you are having to shout to make yourself heard over background noise from the site, then the background noise is likely to be about 75–80 dB.

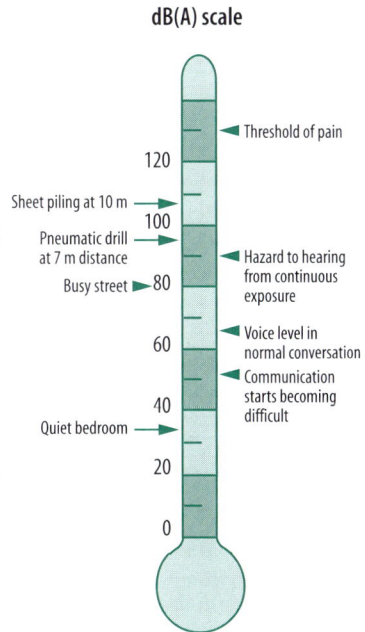

Threshold of pain

120

Sheet piling at 10 m
100
Pneumatic drill
at 7 m distance
Hazard to hearing
from continuous
exposure
Busy street ▶ 80

Voice level in
60 normal conversation
Communication
starts becoming
40 difficult

Quiet bedroom

20

0

Noise units

Noise can be expressed in a number of different ways. The unit dB(A) means that the measured level is in decibels and that it has been passed through a filter so that it represents a level of noise that is very close to a human ear's perception.

In many practical situations (eg a construction site), the noise level varies over time. The L_{Aeq} is the constant sound level which, if it persisted over the same time period, would have the same energy as the varying sound.

Noise propagation

Noise will generally radiate in all directions from a construction noise source, and will bend around and over walls and buildings. It will also reflect back from solid surfaces. Some plant and activities generate more noise in one direction than another, so careful orientation can pay dividends. Screening between the source and a receiver is effective if it obscures the direct line of sight between the two (see page 62).

Noise and vibration

Why noise is important

Excessive noise levels on site represent a major hazard to site workers and can annoy neighbours. Noise causes more off-site complaints than any other topic and can rapidly sour relations. Noise can also disturb our wildlife and natural heritage.

Various types of control on noise levels from construction sites can be imposed when noise starts to cause a nuisance. These controls can affect the programme by limiting the length of time during which noisy activities are allowed and influencing the construction method. Failing to meet noise constraints can result in fines.

If the local environmental health officer considers the noise excessive, a notice to control noise levels or even an injunction to stop work may be served. Some contractors have faced considerable difficulties when their noise control measures have not proved satisfactory to the local authority and have been subject to statutory powers (see Legislation and Standards on page 70).

In some cases residents have taken out civil action against contractors on urban projects.

Contractors are recommended to manage noise in a pro-active way rather than wait for complaints to be made.

Early bird fined

A housing contractor was fined £1000 by magistrates following complaints by local residents about a noisy housing site where work started at 6.30 each morning.

Silence in court

A judge halted work on a £3 million project because the noise was interfering with court proceedings. The judge warned the contractor it would have to pay legal costs if the jury had to be dismissed because they could not hear.

Effects on working hours

On some projects there are agreements on the hours during which noisy working is allowed. On other projects working hours may be restricted by a limit on the average noise allowed to be generated over a given period (either formalised through Control of Pollution Act or COPA notices, under contract conditions, or set out in informal arrangements). In these cases, if the noise exceeds the limit set then the working period <u>must</u> be reduced.

As an example, the actual working time allowed every hour to meet a one-hour average noise level of 75 dB(A) (a typical standard) with different working noise levels is shown below.

Working noise levels dB(A)	Average noise allowed	Actual working time allowed
75	75	60 minutes
78	75	30 minutes
81	75	15 minutes
85	75	6 minutes

Thus, as the working noise level increases, the actual working time allowed at that noise level must be reduced considerably to comply. BS 5228 describes the detailed calculations needed to work out the noise levels caused by a particular project; further guidance is given in CIRIA Project Report 70, *How much noise do you make?* (1999).

Managing noise on site

After complying with contractual noise limits the priority should be to avoid causing conflict with the local community. This will maintain good public relations and reduce the risk of constraints being placed on working practices through the imposition of statutory notices by the local authority.

There are two ways to avoid noise problems on site:

1. to reduce the level of the noise in the community, and

2. to increase the tolerance of those subjected to noise.

To manage noise effectively, use both approaches, as outlined below.

Ensure that noise control features on plant are used. Always keep doors and hoods closed.

Noise and vibration

Reducing noise levels in the community

There are three factors that influence noise levels at a given point:

1. Site management and construction method

2. Plant

3. Screening.

For each of these factors simply employing good practice can bring great benefits. Further reductions require attention to be directed to specific equipment or methods.

In planning the approach to noise reduction on a project, the benefits to be gained from each factor should be weighed against the cost of implementation. In some situations there may be only one solution. BS 5228 shows how to work out the noise from construction operations based on the methods, plant and screening used. The following gives guidance on what to consider for each factor.

Site management and construction method (including timing, duration and phasing)

The general operation of the site needs to be addressed to control noise. It is not only loud noises that cause complaint, but also antisocial activity and irregular or tonal noises such as reversing warnings. Other reasons for complaint include shouting, bad language, radios and out-of-hours deliveries.

Some of the construction activities that cause the greatest problems are: piling (particularly by diesel hammer), breaking out with pneumatic tools, falling ball demolition, earthmoving, scabbling, concrete pours and maintenance works. Calculating noise levels for real operations involves combining the cumulative effects of many different items of plant (see CIRIA PR70 for examples of how this is done). Use the checklist below to minimise noise and vibration.

Checklist – duty of care

- *Change the working method to use equipment or modes of operation that produce less noise. For example, in demolition works can hydraulic shears be used in place of hydraulic impact breakers? In driving steel sheet piles, would the ground conditions suit the jacking method (ie cohesive soils), which produce only a fraction of the noise of conventional hammer-driven piling? When breaking out pavements can methods other than pneumatic breakers and drills be used. Consider chemical splitters or falling weight breakers.*
- *Reduce the need for noisy assembly practices, eg fabricate off site.*
- *Keep noisy plant as far away as possible from public areas.*
- *Adopt working hours to restrict noisy activities to certain periods of the day.*
- *Arrange delivery times to suit the area – daytime for residential areas, perhaps night time for inner-city areas.*
- *Route construction vehicles to take account of the need to reduce noise and vibration.*
- *Keep haul roads well maintained.*
- *Use mufflers or silencers to reduce noise transmitted along pipes and ducts.*
- *Minimise the drop height into hoppers, lorries or other plant (reducing the drop height by a factor of 10 reduces noise by about 10 dB).*
- *Consider using rubber linings on tippers in very sensitive sites.*
- *Liaise with nature conservation bodies to minimise noise disturbance (disruption) to any sensitive wildlife.*

Plant

Noise levels from individual items of plant can vary considerably depending on how they are configured and used. Therefore careful selection of plant is essential when noise is important. BS 5228 provides guidance on noise levels from construction and shows noise levels from typical plant with and without sound control measures. To minimise the noise from the plant on your site observe the following rules:

- use only plant conforming with relevant standards and directives on emissions (see Legislation section, page 70). Older plant, although still legal to use, may not have such identification; as it may be noisier than modern plant, avoid using it in noise-sensitive areas.

Noise and vibration

- When operating plant, use noise-control equipment such as jackets on pneumatic drills, covers on compressors, shrouds on piling rigs and cranes. If in doubt about what is appropriate ask the manufacturer. Hoods and doors on compressors and cranes etc should not only be closed but also be tightly fitting and well sealed. A partly closed door is of little use. Consider placing additional screening around the plant such as plywood screens (see Screening section below)
- electrically powered plant is quieter than diesel- or petrol-driven plant
- operate plant properly so that it does not cause excessive noise. Shut down plant when it is not in use
- maintain plant properly – adequate lubrication to reduce squeaks and the tightening of loose nuts and bolts to minimise rattles are part of routine maintenance
- provide effective silencers for plant, eg pneumatic percussive tools
- fix rotating or impacting machines on anti-vibration mountings
- ensure that audible warning systems (including reversing alarms) are switched to the minimum setting required by the Health and Safety Executive. Consider the use of alternative systems (eg cab-mounted CCTV) where appropriate. Traffic routes that avoid reversing on site will minimise the impact. Use tannoy systems only when necessary.

Be aware that noise from some plant such as compressors may be emitted more in some directions than others. Therefore consider the orientation of static plant. If you put plant next to a solid surface (eg a wall) the noise will be reflected away from that surface and increased by 3 dB(A).

Screening

If designed and used correctly screens can reduce noise levels from a site considerably at relatively low cost. Factors affecting the efficiency of a screen include: distance from the source and from the receiver of noise, density of material used, height and length, the presence of holes and its position relative to noise-reflecting surfaces.

Noise screens can consist of topographical features as well as artificial materials (but trees do little to reduce noise). To be effective an artificial screen should have a density of at least 7 kg/m^2 of surface area. This equates to 12 mm thickness of plywood. Site hoarding can therefore have a screening effect, although its value is diminished if it is distant from the source and receptor. If hoarding is acting as a noise screen, worksite gates should be opened for the minimum amount of time to allow the passage of vehicles.

Sound rules

In an open field, if you double the distance between the receptors and the source you reduce the noise level by 6 dB. This reduction will be less in built-up areas.

Halving the amount of plant or working time reduces the noise level by 3 dB.

To add two noise levels together use the following table:

Difference between levels in dB	0	1	2	3	4	5	6	7	8	9	10
Add to higher noise level to get resultant noise	3	2.5	2	2	1.5	1	1	1	0.5	0.5	0

For example, one piece of plant is operating on site with a noise level of 74 dB, and an additional piece of plant with a noise level of 70 dB is brought in. The difference between the two noise levels is 4 dB, so using the table shown above you add 1.5 dB to the higher noise level. Thus the noise level resulting from bringing in the extra piece of plant is 75.5 dB. If the additional piece of plant operated with a noise level of 60 dB rather than 70 dB (as above) then the resultant noise level would remain at 74 dB.

Case Study

On one site, temporary lighting was installed at a road diversion near residential properties. To reduce the noise, the contractor placed hay bales at a safe distance around the generator.

Where sensitive receivers of noise are on one side of a site, a three-sided enclosure with an opening on the side facing away from the sensitive area can be effective. If there is a barrier (eg site hoarding parallel to the enclosure), cover the enclosure's internal surfaces with a sound-absorbing material such as mineral wool quilt to prevent the build-up of reverberate sound. The figure below shows an easily constructed, effective, mobile screen.

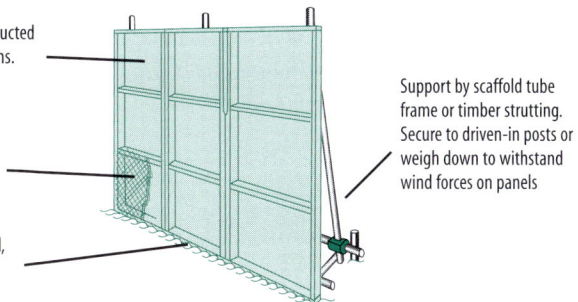

Screen made up of formwork panels, or constructed from at least 12 mm thick plywood and battens. Ply may need to be stiffened with additional battens to prevent drumming.

Fibrous material held in place by wire mesh will improve performance if placed on side facing noise source.

Lower edge of panels must rest on the ground, and any gap plugged by spoil, sand bags etc.

Support by scaffold tube frame or timber strutting. Secure to driven-in posts or weigh down to withstand wind forces on panels

Noise and vibration

Example of how screening can reduce or attenuate noise

The higher a screen is, the more effective it is. A screen that is placed near to either the noise source or the receptor is more effective than one which is placed in the middle of the two. These points are demonstrated in the example below.

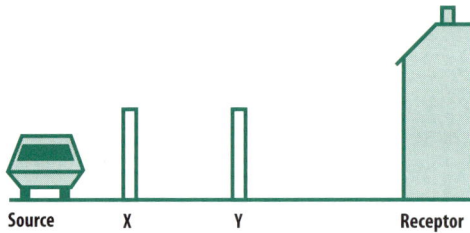

The source is 100 m from the receptor. At 10 m away from the source, the reduction in noise at the receptor from a 2 m-high screen placed at X is 8 dB(A). With the screen located at mid distance (Y), the reduction in noise is only 6 dB(A).

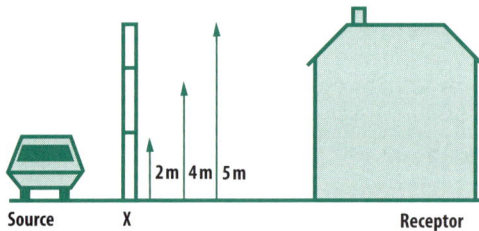

With a 2 m-high screen placed at X, the reduction in noise at the receptor is 8 dB(A). If this screen is 4 m high the reduction is 13 dB(A), and if it is increased to 5 m, it brings a reduction of 15 dB(A).

Checklist – design and placement of screens

- Where possible, place sources of noise away from sensitive areas.
- Avoid sound-traps that amplify noise.
- Almost any solidly built screen is better than none.
- Erect the screen close to the source of noise.
- Build the screen from stout materials, with panels stiffened to prevent drumming.
- For the most effective results build the screen about 1 m above the highest sight line.
- Seal all gaps and openings, including gaps at the bottom of the screen.
- Glaze any public observation openings in perimeter hoardings with Perspex (protected with wire mesh or similar) if sensitive areas are closer than the height of the hoarding.
- Consider placing additional screens close to sensitive areas but not parallel to nearby walls.

Increasing community tolerance

Community liaison is the key to increasing community tolerance to noise. If you inform the local community and residents of what you will be doing on site and for how long, they may accept higher noise levels. Discuss with them in advance what aspects of noisy working is most annoying and see if there are any remedies that will not affect your programme. Consider agreeing to a shutdown at particularly sensitive times, for example halt noisy works at lunchtimes to appease restaurant- and pub-owners. Through good relations, the potential for complaints or civil claims in the long term may be reduced. It is especially important to avoid unexpected early starts in the morning.

Case Study

During the construction of the Docklands Link Road many residents were affected by noise. Depending on the predicted noise levels, residents were offered new windows to a high sound insulation specification, and some were even given the opportunity to relocate temporarily. Where residents chose to remain at home, cash compensation payments were made to the worst-affected.

Case Study

An inner-city site manager needed to phase noisy works without disturbing nearby office workers. He even went to the extent of stopping all noisy works during his neighbours' key meetings, which were held three times a year.

Noise monitoring

The need to undertake noise monitoring may be an obligation of the contract or may arise out of a noise agreement with the local authority (eg Section 60 or Section 61 notice under the Control of Pollution Act, see page 70). Although COPA is mainly used in relation to construction site noise, the provisions of the Environmental Protection Act (EPA) must not be ignored (see page 70).

Even if monitoring is not obligatory it is often useful to obtain a series of background noise levels before construction operations begin. However, short lead-in times between contract award and start on site may limit the time available for such background monitoring.

When construction work starts and unless otherwise specified, readings should be taken on a daily basis for the first week to verify that noise levels are reasonable and not in contravention of any limits that have been imposed.

Noise and vibration

Subject to satisfactory results, the frequency of measurements can then be reduced to a survey once a week, although the frequency should be increased again if particularly noisy activities (such as driven piling) are undertaken. The duration of the monitoring on any one day is important. BS 5228 suggests that the average noise level over a day can be approximated within certain tolerances by taking short-term measurements throughout the day, eg 5 minutes in every hour.

If noise monitoring is not carried out by a suitably trained and experienced person the results may be misleading or inaccurate. The following information is given to help site personnel supervise whether monitoring is being done properly.

How to measure

BS EN 60651 : 1994(13) classifies sound-level meters according to the precision (or tolerance) to which they measure. The highest-precision meters are classified Type 0, whereas the lowest grade are Type 3. It is often the quality of the microphone (the sound transducer) that dictates the grade of performance of the meter. Specifications usually call for the use of a Type 1 or 2 meter on construction sites.

To measure LAeq values an integrating-averaging sound-level meter should be used. The relevant standard is BS EN 60804 : 1994(14), and precision is graded as for normal sound-level meters.

Whatever recording system is adopted, calibrate the meter before and after measurement. Protect the microphone with a foam windshield; where extended measurements are to be taken in poor weather conditions consider using an all-weather microphone protection enclosure. Although it is possible to hold the meter in the hand for spot checks, it is preferable to mount the meter on a suitable tripod.

Where to measure

The location of measurements may be imposed (by previous agreement, eg 1 m from the exposed window of the nearest noise-sensitive dwelling or premises, or "at the site boundary" or "1 m from the site boundary".

Establish whether the measurements are to be made in a free-field condition or near a reflecting surface. As a guide, for free-field measurements the microphone should be at least 3.5 m away from reflecting surfaces.

For long-term and possibly unmanned monitoring, take extra care in locating the measurement equipment in order to minimise the risk of damage, theft and the influence of other, external noise sources.

Why vibration is important

Although rare, high vibration levels over sustained periods can cause damage to buildings and sensitive equipment within buildings, such as computers. Lower levels can cause nuisance to residents; the degree of annoyance depends on the activity, the persons affected and the vibration intensity. It is likely that local residents will complain about any perceived vibrations as soon as they become noticeable.

Vibration may also cause disruption to wildlife, and damage to geological, geomorphological and archaeological sites. The level at which this occurs is highly site-specific.

Since the effects of high-frequency vibration are less than those of low-frequency ones, it is worth seeking ways to change the frequency if a problem is being caused. This usually requires specialist advice; ask plant manufacturers for their help.

How to avoid vibration problems

There are three primary aims in the management of vibration on site:

- to avoid causing damage to nearby structures
- to avoid causing annoyance and concerns
- to avoid being falsely accused of causing damage.

The following six steps will help in addressing each of these aims.

STEP 1 *Evaluate the potential for vibration and thereby damage*

It is not possible to give simple guidance on the vibration levels that would arise at nearby properties from various types of plant. This is because transmission of vibration is highly dependent on ground conditions and on features in the ground such as pipe runs. Although prediction methods exist, they require specialist expertise to implement and rely on detailed information being available on the site, the ground conditions and the plant.

There is no UK standard or code giving guidelines or assessing the risks to buildings from ground-borne vibrations. BRE Digest 353 refers to German, Swiss, Swedish and draft ISO standards, but points out that UK conditions and structures may differ from those on which these guidelines are based. It is best, therefore, to consider each situation separately. If it appears that a piling or ground engineering project is likely to generate vibrations approaching the intensities considered as putting adjacent structures at risk (using the general limits given in those international documents or local knowledge), then a detailed study should be undertaken by specialists. In most situations, however, it should be sufficient for staff to review the operations to take place on site and establish the sensitive areas around the site that may need to be monitored.

Noise and vibration

STEP 2 *Monitor conditions before works start*

Before starting the construction operations, it is important to survey sensitive locations and structures. The survey should include a detailed record of:

- existing cracks and their widths
- level and plumb survey, including damp-proof course
- measurements of tilting walls or bulges
- other existing damage including loose or broken tiles, pipes, gullies or plaster.

Photographic records and the installation of measurable tell-tale devices are also helpful to establish alleged or actual damage. In some situations it will be necessary to strengthen vulnerable off-site structures before vibrations start.

Sensitive locations to survey and monitor before and during construction would include:

- schools
- hospitals and nursing homes
- historic buildings
- museums
- laboratories
- precision machine workshops
- sensitive plant or equipment used by local companies
- housing
- buildings in poor condition
- brittle/ancient underground services including tunnels
- geological, geomorphological or archaeological sites, and certain wildlife.

STEP 3 *Inform neighbours*

Vibration causes anxiety and annoyance to residents mostly because they fear that it will cause damage. It is therefore useful to explain to them that damage only occurs at vibration levels many times greater than those that can be felt from construction plant.

Informing neighbours of the potential for vibration allows the site staff to learn of any particularly sensitive issues that may be time-dependent and that may be resolved by limiting hours of work.

Manufacturers should be able to advise on the level of vibration that might harm computer installations.

STEP 4 *Minimise effects during works*

Reducing vibrations during the works is difficult to achieve because they are a fundamental side-effect of the process being undertaken.

To reduce vibrations the methods being proposed will need to be re-evaluated. For example, piling is well known for causing vibration effects, but driven casings cause greater effects than vibrated casings. When evaluating how to reduce vibrations consider the following:

- high-frequency vibration causes less damage than low-frequency vibration
- isolating plant from the transfer medium is an effective solution
- plant placed on a heavy base cause less vibration than plant on a lighter base, eg suspended slabs
- vibrations travel less distance in unsaturated ground. If groundwater levels fluctuate (eg in tidal regions) carry out works during lower tides.

STEP 5 *Monitor vibration levels during the works*

To be effective, vibration-level monitoring needs to be carried out by trained staff or by external specialists. However, it may be necessary for site staff to discuss with building occupants where vibration monitoring can be conducted. The two main rules are that when monitoring at properties during operations:

- measure inside rooms when assessing for nuisance
- measure on the structure outside when assessing for damage (doorsteps are a good location).

Do not forget that in sensitive structures continued visual monitoring and measurement of crack widths is the best way to determine whether damage is being caused.

STEP 6 *Monitor conditions after works are completed*

The same evaluation as undertaken for Step 2 should be carried out and results compared to the pre-condition surveys.

Key Guidance

- *How much noise do you make? A guide to assessing and managing noise on construction sites, CIRIA Project Report 70, 1999*
- *BS 5228 (1997) Noise control on construction and open sites*
- *Noise and vibration from piling operations – CIRIA PG9, 1980*
- *Planning to reduce noise exposure in construction – CIRIA Technical Note 138, 1990*
- *Ground-borne vibrations arising from piling – CIRIA Technical Note 142, 1992*
- *Noise Control: Principles and Practice – Bruel and Kjaer (Naerum, Denmark), 1982*
- *A Guide to Reducing the Exposure of Construction Workers to Noise – CIRIA Report 120, 1990*
- *Simple noise screens for site use – CIRIA Special Publication 38, 1985*

Noise and vibration

Legislation and standards

Environmental Protection Act 1990

The Environmental Protection Act 1990 (EPA) contains a number of provisions in Part III relating to noise as a statutory nuisance. Sections 80 and 82 provide for proceedings to be instigated by the local authority (Section 80) or by an individual (Section 82) where either considers noise (or any of a number of other "pollutants") to constitute a nuisance. Under Section 80 the local authority may serve an abatement notice; under Section 82 an individual may obtain an abatement order through a court.

Some local authorities have been known to invoke a Section 80 abatement notice, with potentially more severe consequences than those normally associated with a Section 60 notice under Control of Pollution Act. In addition, any person is able to seek an injunction in a civil court: under common law, noise has long been recognised as a nuisance by "interfering with the enjoyment of one's premises".

Control of Pollution Act 1974

A local authority can control noise from construction sites by serving a Section 60 notice, which specifies what plant and machinery may be used, the working hours and noise levels. A developer may apply for prior consent for construction works through a Section 61 consent. It is a defence against any prosecution under Section 60 to have in place a Section 61 consent provided that the developer is complying with the terms and conditions of that consent. It is also a defence, whether or not a Section 61 consent exists, to demonstrate that best practicable means are being used to minimise noise emissions. "Best practicable means" (BPM) is defined in Section 72 as having regard to, among other things, local conditions and circumstances, the current state of technical knowledge and reasonable cost. The local authority can also designate noise abatement zones in which specified types of development may not exceed specified noise levels. Since either Section 60 or Section 61 procedures may be appropriate, designated site staff should be familiar with the requirements of each.

In Northern Ireland, similar provisions are made to Section 60/61 in the Pollution Control and Local Government (Northern Ireland) Order 1978.

Noise at Work Regulations 1989

This is a statutory legal document that details provisions for the protection of people's hearing while in the workplace. Various levels of acoustic exposure are specified which, when exceeded, determine the level of protection required. Failure to comply with the specifications may lead to prosecution under the Health and Safety Regulations.

Noise and vibration

Construction Plant and Equipment (Harmonisation of Noise Emission Standards) Regulations 1985 and 1988

An EC type-examination certificate is required before any item of construction plant and equipment may be marketed. Construction plant and equipment must carry an EC mark to indicate that it conforms to the levels given in the Regulations for that type of machinery. Failure to comply with, or contravention of, the Regulations may result in a fine of up to £2000.

BS 5228 – Noise control on construction and open sites

As the nature of work on construction sites varies so enormously, regulation to control the noise from such activities requires some flexibility. This standard attempts to do this. It comprises four parts. Part 1 gives basic information and procedures, Part 2 deals with construction and demolition, including road maintenance, Part 3 applies to coal extraction by opencast methods and Part 4 deals with piling operations. The definition of acceptable noise levels is outside the scope of BS 5228.

Noise standards and guidelines

The local authority has the power to fix limits under the Control of Pollution Act and the Environmental Protection Act. In the absence of local authority guidance, the values given below are a useful guide. Daytime levels outside the nearest window of the receiver's dwelling should not exceed:

- 70 dB(A) in rural, suburban and urban areas away from main roads and industrial noise
- 75 dB(A) in urban areas near main roads and heavy industrial areas.

These levels recognise the temporary nature of the operations. However, they should not be considered a desirable level because, in summer, for example, the attenuation from outside to inside a house with open windows is only approximate 10 dB. This would result in levels of 60–65 dB(A) inside the dwelling.

For evening operations BS 5228 suggests that noise limit should be 10 dB(A) below the daytime limit, but recommends that the average steady sound levels as low as 40–45 dB(A) may be necessary at the facade of noise-sensitive premises to avoid sleep disturbances.

BS 6472 (Evaluation of human exposure to vibration in buildings)

As a general guide to acceptable limits for vibration in buildings, the guidance in BS 6472 suggests that levels of vibration below 0.1–0.2 mm/s will normally be imperceptible to people in most situations. This should be the target level. In exceptional circumstances, higher levels may be allowable provided the levels do not exceed those deemed to be limiting values for damage to buildings.

Dust, emissions and odours

Why dust, emissions and odours are important

Dust, emissions and odours arising from a site will annoy neighbours and can even cause health risks at very high concentrations. There is also the potential for legal action, which will have cost and programme implications. Dust, emissions and odours can be particularly hazardous to site staff in confined spaces, so seek information on controls from the company health and safety officers.

> Dust is generally considered to be any airborne solid matter up to about 2 mm in size. Particle sizes can vary considerably, depending on their origin, and the smallest particles can be breathed in. Some dust, such as limestone dust, is chemically active.

Annoyance to neighbours

Dust, emissions and odours disturb site neighbours. Annoyance is caused when residents have to re-clean washing that has been hanging out and when they have to wash cars, curtains and windows. Windblown dust can be unsightly over long distances in scenic areas. In exceptional circumstances, dust can affect health by, for example, causing eye irritation. Asthma can be exacerbated by exposure to respirable dust.

Claims from farmers for dust damage to crops

Claims are particularly common on rural road projects. Even very low concentrations of dust can affect plant and fruit growth. Plant growth is especially susceptible to dusts that are highly alkaline, for example limestone, and cement dust. The extent of effects caused is weather-dependent. For example, dust deposited during light rain can cause the surface of the soil to form a crust. Claims for damage to crops in excess of half a mile from the site have been made because dust can be blown for long distances.

Impact on project programme and budget

Some contracts may require contractors not to work at times of high wind in a certain direction. Working to comply with strict dust levels can impose cost and/or programme constraints (see case study). If a statutory nuisance is caused (see page 79) an abatement notice may be served.

Impacts on ecology

Dust blowing onto watercourses can damage the ecology, particularly if the watercourse is sensitive. As identified above, dust may also affect plant growth and alkaline dusts may change species composition in some situations (see case study). This may particularly affect new growth planted as part of the contract. Ash trees may drop their leaves up to six to eight weeks early following exposure to high levels of dust.

Impact on plant and equipment

Within the site, dust can cause mechanical or electrical faults to equipment such as computers and will increase abrasion of moving parts in plant and clogging of filters.

Case Study

Topley Pike is a Derbyshire heathland that has been affected by limestone dust deposition. The unpolluted heaths have a soil pH of 3–4 and are dominated by plants that are adapted to relatively acidic environments. The dust-affected areas have a soil pH 1.5 units higher, ie less acidic, resulting in domination by other species, and several tree species may have died as a result. The consequence of this change in the vegetation has been an alteration in the animal community, with herbivorous insects and the animals that feed on them also declining in numbers.

A company was fined £8500 under air pollution legislation for twice carrying dusty material in uncovered containers for a short distance on a public road.

How to avoid problems

With dust, emissions and odour there are usually no imposed standards to be met for a particular contract. Regulators become involved only once problems have been created and complaints received. To avoid causing complaints, the site should operate a management system that ensures that:

- dust, emissions and odour from general operations are minimised through adoption of good working practice
- special consideration for control measures is given in circumstances where general good practice may not be sufficient to avoid causing problems.

It is also valuable to keep a record of daily dust conditions in case disputes arise (see also page 78).

If annoyance is still caused following implementation of dust-control measures, find out the reason and offer to put it right by, for example, cleaning windows or washing affected cars.

Dust suppression

As it is difficult to suppress dust once it is airborne, it is essential to develop a strategy to stop dust being generated. Careful design of construction operations, including the location of stockpiles and batching plant, can reduce dust. Refer to the checklist on the following page.

Dust, emissions and odours

Checklist – avoiding dust generation

Haul routes

- Select suitable haul routes away from sensitive sites if possible.
- Pave heavily used areas, or use geotextiles eg around batching plant or haul routes. Sweep these regularly.
- Provide a length of paved road before the exit from the site.
- Reduce the width of haul roads (while still allowing two-way traffic) to minimise surface area from which dust may be produced.
- Sweep paved access roads (whilst still allowing two-way traffic) and public roads regularly using a vacuum sweeper.
- Limit vehicle speeds – the slower the vehicles the less the dust generation.
- Damp down – see opposite.

Demolition

- Use enclosed chutes for dropping to ground level demolition materials that have the potential to cause dust and regularly dampen the chutes.
- The use of mobile plant for crushing materials such as bricks, tiles and concrete is covered by the EPA 1990 (see Key Guidance).
- Locate crushing plant away from sensitive sites – consider siting within buildings (eg buildings within the site that will not be demolished or those to be demolished last).

Plant

- Clean the wheels of vehicles leaving the site so that mud is not spread on surrounding roads – dry mud turns to dust.
- Ensure that exhausts do not discharge directly at the ground.

Earthworks and excavations

- Revegetate or seal temporary or completed earthworks as soon as possible.
- Keep earthworks damp – try to programme to avoid exceptionally dry weather.

Materials handling and storage

- Locate stockpiles out of the wind (or provide wind breaks) to minimise the potential for dust generation.
- Keep the stockpiles to the minimum practicable height and use gentle slopes.
- Compact and bind stockpile surfaces (in extreme cases). Revegetate long-term stockpiles.
- Minimise the storage time of materials on site.
- Store materials away from the site boundary and downwind of sensitive areas.
- Ensure that all dust-generating materials transported to and from site are covered by tarpaulin.
- Minimise the height of fall of materials.
- Avoid spillage, and clean up as soon as possible.
- Damp down – see box opposite.

Concrete batching and pouring

- Mix large quantities of concrete or bentonite slurries in enclosed/shielded areas.
- Before concrete pours, vacuum dirt in formwork rather than blowing it out.
- Keep large concrete pours clean after they have gone off. They generate large quantities of dust.

Cutting/grinding/grouting/packing

- Minimise cutting and grinding on site.
- On cutters and saws, use equipment and techniques such as dust extractors to minimise dust. Consider a wet cutting saw or use vacuum extraction.
- Spray water during cutting of paving slabs to minimise dust.

Damp down using water

The most effective application of water in suppressing dust is by using a fine spray, but the efficiency depends on the speed of the bowser. Repeat spray regularly and frequently, especially during warm and sunny weather when water will evaporate quickly. Consider spraying:

- unpaved work areas subject to traffic or wind
- structures and buildings during demolition
- sand, spoil and aggregate stockpiles (this has only a very temporary and slight effect)
- during loading and unloading of dust-generating materials.

If you are abstracting water from a watercourse, ensure that you have obtained permission (see Water 3.1).

Damp down using water with chemical additives or binders

Although spraying with water is effective when it is first applied the effect may not last long. Repeated application, particularly in drought seasons, may itself be environmentally unacceptable. For some applications, and particularly for haul roads, the addition of a water-retaining binder may be beneficial. Spraying with water and chemical additives is more effective than using water alone because it reduces the number of passes per day and the volume of water needed. Several proprietary methods are available:

- calcium chloride – spread on unmade aggregate roads: neat at 2 kg/m^3 or made surfaces at 1.5 liquor/water dilutions
- magnesium chloride – claimed to be 95% effective on sand and gravel roads
- proprietary brands – especially polymer bonding agents applied as a water additive.

These have been used successfully in the past, especially at mineral workings.

There is no general guidance as to which additive is best. The cost of the additives has to be weighed up against savings in water supply, bowser usage and downtime. Take care to avoid over-application, which may cause pollution.

Seek advice from the Environmental Agencies before using additives.

Dust, emissions and odours

Effective planning and management of dust control requires a knowledge of wind conditions for the site. Although prevailing winds across most of the country are from the south-west, there are regional and seasonal variations. For example, in winter months winds are often from the north-east. General historical wind data for the site locality can provide guidance on the likely wind speed and direction. This can be obtained from the Meteorological Office in Bracknell in the form of wind roses.

If dust is likely to be an important issue it may be necessary to plan future works against a short-term weather forecast. Obtain this from the local Meteorological Office. Such information must be used with care because wind directions and speeds on some sites are significantly influenced by local landforms and building features. Local people may be able to provide guidance on local patterns.

Dust screening

If dust-generating activities cannot be avoided, it may help to erect screens to act either as windbreaks or as dust screens. These can take the form of permeable or semi-permeable fences, but be aware that they can be expensive if designed to resist high winds. Trees or shrubs planted early as part of site landscaping can provide some screening; likewise retention of existing vegetation (or buildings to be demolished) will aid screening.

Case Study

On one site, screening in the form of transparent sheeting was provided alongside bridges over sensitive watercourses.

Emissions and odours

Processes involving the use of fuels and the heating and drying of materials commonly emit fumes, odours or smoke. It is important to prevent emissions and odours as far as possible, to protect workers and because they annoy the public and affect the environment. Take the preventative measures listed in the checklist below. Any works that involve the risk of creating odours (eg works on sewers) should be phased carefully.

Checklist – preventing emissions and odours

Vehicles and plant

- Keep vehicles and plant used on site well maintained and regularly serviced.
- Ensure that all vehicles used by contractors comply with MOT emissions standards at all times.
- Control deliveries to site, to minimise queuing.
- Make sure that engines are switched off when they are not in use. (This is particularly important in summer near buildings that do not have air conditioning.)
- Control staff car parking to minimise queuing.
- Keep refuelling areas away from the public.

No fires on site

- Do not burn waste materials/tyres on site.

Waste storage

- Use covered containers for organic waste and remove frequently.
- Remove organic waste (eg weeds and other vegetation) before it begins to decompose.

Chemicals on site

- Store fuels and chemicals and other dangerous substances in the appropriate manner.
- Take account of the wind conditions when arranging activities that are likely to emit aerosols, fumes, odours and smoke.
- Position site toilets away from public areas.

Dust, emissions and odours

Dust limits

There are no nationally accepted criteria defining levels of dust that either cause a nuisance or potential health risks. There are, however, general guidelines that relate to either the level of deposition of dust (measured as milligrams per square metre per day or in terms of the percentage coverage of a surface area per day) or the level of dust in suspension in the air (measured as micrograms per cubic metre). Levels of acceptability can also relate to the percentage increase in the background deposition rate, which itself can vary substantially.

Dust prediction and monitoring

Whether or not quantitative dust monitoring is imposed on a project, contractors should keep a daily log. This should note weather conditions, construction activities, their location on site and visible dust-generating activities. If a complaint is made, the contractor can refer back to the log to see what was happening on site that day. A photographic record may also be useful especially if it records the dust-control measures employed.

If monitoring is imposed, there are two main methods that can be used on site:

1. Exposing microscope slides or sticky pads for a given period and calculating the deposition rate of dust over the exposed period.

2. Using high-volume samplers to draw air through a filter to measure the volume of dust in the air at the time.

Neither method provides definitive evidence of the dust impact and both may be expensive. Experience has shown that much of the dust collected on slides or pads does not arise from the construction site. Nevertheless, it may be useful for a contractor to volunteer to monitor dust as a demonstration of its commitment to good practice.

For a dust-monitoring programme to give more definitive evidence of construction impacts it should include both upwind and downwind monitoring of the site and cover a baseline period before construction started. (The baseline period should ideally cover the same seasons as the construction period.) This level of monitoring is not generally recommended unless required by the contract or if the impact of dust is expected to be significant.

Computer models are available for predicting levels of dust generation and movement. However, they require a considerable level of detail of operations, particle sizes and weather conditions (wind and rainfall) and do not give precise results. They are therefore not recommended for general site use.

Dust, emissions and odours

Key Guidance

Secretary of State's Guidance – Mobile Crushing and Screening Processes, 1996, Department of Environment, PG3/16(96).

Environmental impacts of surface mineral workings, DETR, HMSO.

Legislation

Environmental Protection Act 1990 – statutory nuisance provision

Dust, emissions and odours often generate complaints of discomfort or inconvenience. This may constitute a nuisance under statutory law where the wellbeing and personal comfort of residents and the use and enjoyment of their property is being affected. The problem does not need to cause harm to health, but it must cause significant inconvenience to be considered a nuisance.

The person responsible (in practice the contractor doing the work in question) can be required by the local authority to put a stop to a statutory nuisance by serving an abatement notice. Aggrieved individuals can also apply to the magistrates for an abatement order to stop the nuisance. Breach of an abatement notice or order is a criminal offence.

Clean Air Act 1993

This is the primary legislative means of control over smoke, grit and dust. It is generally enforced by local authorities, and applies to emissions of dark smoke from industrial or trade premises that originate from sources other than chimneys. Demolition has been found to be a trade process for the purposes of this legislation and a bonfire on a demolition site emitting dark smoke has been found to be within the ambit of the Act. However, there are regulations that exempt emissions from the burning of timber and most other waste resulting from demolition of a building or of a site and also emissions from tar, pitch, asphalt or other matter used in connection with surfacing.

Ground contamination

If contamination is likely to be encountered the contract should define the methods of dealing with it. The contract will usually refer to the guidance issued by statutory authorities on how to deal with contamination. This type of detailed advice is beyond the scope of this handbook – look for further advice on page 84. This section advises on how to deal with contamination (both expected and unexpected) and how to avoid causing or spreading contamination.

Why is contamination important?

Ground contamination may result in the following problems:

- health and safety impacts on staff and surrounding community through exposure to contaminants
- liability for the cost of disposal or remediation of contamination (depending on contract conditions)
- liability for costs arising from unexpected spreading or making existing contamination worse
- delays to the programme through unexpected or accidental contamination
- pollution of groundwater and surface water courses
- pollution of surrounding land.

> A major housebuilder and two other defendants were ordered by the Crown Court to clear 270 lorry-loads of waste containing PCBs and oil that had been dug out from a redundant gasholder and illegally dumped on farmland. The success of the operation was important because the defendants were told that the level of fine would depend upon the clean-up achieved.

There are two key requirements in managing ground contamination:

- do not cause it or spread it
- deal with it appropriately.

The main causes of problems

Generally, ground contamination is present as a result of previous uses of the site (or adjacent land), for example industrial land use or waste disposal. Be aware that incidents during construction may also cause ground contamination – as shown below.

Examples of how problems with contamination may arise

Existing contamination

- encountering contamination during the works
- handling or excavating contaminated ground – for example where impermeable ground is penetrated and therefore creates a pathway to an aquifer below.

Causing or spreading contamination

- windblown contaminated dust arising from loading of lorries and transportation
- stockpiling contaminated ground on clean ground in the course of works on site
- stockpiling materials containing contaminants that are liable to leach out
- spillages of contaminants such as oil onto the ground during construction
- dewatering that draws in contaminated groundwater from adjacent sites
- discharge of contaminated dewatering water into nearby watercourses.

Disturbing unexpected archaeological finds, such as burial grounds, may pose a health risk to the workforce.

How to avoid problems

Initially, the responsibility lies with the client for carrying out a site soil survey and deducing the risks from any contamination that may be present. However, the contractor is advised to check whether this has been done satisfactorily. A preliminary investigation should have included an investigation of the site history and its surroundings, involving examination of published maps, plans and photographs and existing site records and enquiries. Information relevant to a preliminary investigation includes:

- the history of the site (details of its owners, occupiers and users)
- the processes used (including their locations, raw materials, products, waste residues and methods of disposal
- the layout of the site above and below ground at each stage of development (including roadways, storage areas and other hard cover areas)
- the presence of waste disposal tips, made ground, abandoned pits and quarries with or without standing water
- mining history (including shafts and roadways)
- information on geology and hydrogeology (including the presence of groundwater and surface water)
- potential contamination uses of sites, past or present, in the area adjacent to the site.

Ground contamination

If the preliminary investigation finds that contamination is probable, an exploratory investigation is normally undertaken to define the type, concentration and extent of any contamination on the site. If the site manager suspects that contamination might be an issue on the site and is concerned that there is a lack of information, then advice should be sought.

Information from a phase 1 (preliminary) investigation and a phase 2 (exploratory) investigation should have been supplied to the contractor as part of the health and safety plan. The contractor will be required to prepare H&S risk assessments and detailed method statements, which will state the approach to be taken to deal with the contamination on the site.

It is always possible that contamination will not have been located during investigations but will be uncovered during the site works. This is most likely to occur at sites with an industrial history, although old waste tips may be found on any site. Therefore, site staff – especially excavator operators – should be vigilant during excavations (see the checklist of visual signs).

Case Study

A JCB operator was excavating a pipeline in the vicinity of a waste dump. During excavation, a soil slip resulted in part of the dump being exposed, including a number of drums. The JCB operator endeavoured to replace a slipped drum on the dump and in so doing pierced it – exposure to the sun had caused it to expand and so the drum exploded, releasing an unknown gas. The driver was exposed to the vapours released from the drum and had to receive medical attention.

Checklist – avoiding causing or spreading

- Do not stockpile contaminated soil unless it cannot be avoided. If it is necessary, stockpile only on a hardstanding area to prevent contamination of underlying ground. It may be necessary to cover over stockpile material, either to prevent windblown dust or to prevent ingress of rainwater. Control surface drainage from stockpiled area. Water draining from a stockpile may be contaminated and need controlled, off-site, disposal.
- Prevent the spread of contaminated dust – refer to Section 3.4.
- Be careful when handling, storing and using oils and chemicals – refer to Section 4.18.

What to do if contamination is encountered

As mentioned above, the contractor should already have adopted appropriate working methods to deal with contamination that is expected. These will be detailed in the risk assessment. The following section deals with unexpected contamination.

During boring, digging, excavating and similar operations, observe the uncovered ground and watch out for visual signs of contamination. The release of noxious fumes (petrol, oils, solvents, chemical residues) and smells may also indicate contamination (eg a smell of bad eggs may mean that hydrogen sulphide contamination is present).

Checklist – visual signs

- discoloured soil (eg chemical residues)
- fibrous texture to the soil (eg asbestos)
- presence of foreign objects (eg chemical/oil containers)
- evidence of previous soil workings
- evidence of underground structures and tanks
- existence of waste pits
- made ground (ie artificial ground where ground level is raised by man's activities and not due to a natural cause)
- old drain runs and contamination within buildings; tanks, flues etc
- topsoil adjacent to motorways can be contaminated by traffic emissions.

Covering up problems permanently does not get rid of them! When contamination is suspected do the following:

- stop work immediately
- seal off the area
- report the discovery to the site manager
- the site manager should seek expert advice.

> If asbestos is uncovered re-cover it temporarily to prevent its release to the atmosphere.

If there is a risk of spreading contaminated dust, temporarily cover the source or dampen it down.

Ground contamination

Always wear the correct protective clothing

Legislation

Although there is no legislation set out specifically to address contaminated land issues several statutes influence how contaminated land issues are managed, including:

- the Environmental Protection Act (prescribed processes and substances) Regulations 1991 prescribes substances where release into the environment is controlled
- the Environmental Protection Act 1990 (as amended) identifies contaminated land as:
  – land that is causing (or has the potential to cause) significant harm, or
  – land that is polluting (or has the potential to pollute) controlled waters.

Under this legislation, local authorities have the power to issue a remediation notice to specify what action needs to be taken by the person who contaminated the land. Failure to comply with a remediation notice is an offence.

- Waste management legislation (see 3.2)
- health and safety legislation.

What are wildlife and natural features?

In this handbook these terms are taken to mean all living things, such as trees, flowering plants, insects, birds and mammals and the habitats in which they live, as well as the rocks and landforms that shape the landscape. Climate, geology or soil structure and land management all influence the creation of habitats and the presence of species in particular locations. It is therefore important to look not only at individual living things but also at larger animal populations, plant communities, their habitats, the rocks and landforms and the natural processes that affect them.

Why is it important to consider it?

Wildlife and natural features are vital to the quality of life and a healthy environment. They are becoming more valued by the general public and those prepared to demonstrate against construction projects. At the same time the level of protection given to them is increasing – through legal controls and in contract conditions. Usually, the developer has responsibility for investigating the ecological constraints and sensitivities of a site and defining them to the contractor. If contractors fail to meet their legal and contractual requirements, sanctions may be imposed which can affect the cost and the programme for the project. Beyond these requirements, contractors should adopt a responsible attitude by ensuring that their activities cause the least damage to the surrounding natural environment (and, where possible, enhance it).

Be aware!

Unexpected ecological finds can arise during the works. The contractor has the responsibility to deal with these in the correct manner. This may affect the works.

Fines and costs for non-compliance

Damaging or disturbing protected species can result in prosecution under a range of legislation. The fine for non-compliance with legislation varies according to the species and the type of damage caused. For example, if protected species such as bats are disturbed, fines may be imposed at £5000 per animal. There is also scope for the confiscation of any vehicles or other equipment used to commit the offence.

Recently, a developer was fined £25 000 for causing irreparable damage to an oak tree.

Wildlife and natural features

Contract or planning conditions

Contract or planning conditions may state that certain trees must remain undamaged. Replacing damaged mature trees is expensive – a 10 m-high tree may cost £2000, plus extra expense for delivery, planting and several years' maintenance.

Impact on project programme

In many cases, works have to be timed in order to avoid disturbance to species during particularly sensitive times (eg hibernation or mating seasons, and the bird-nesting season). If this is not considered early enough in the project programme, disruptions may be caused.

When damage has occurred, negotiating with the nature conservation bodies and repairing damage takes time and may cause delays to the programme (see box below for explanation of the nature conservation bodies).

Public awareness

Today the public is very aware of impacts to the natural environment and the damage that can be caused. Many projects are under the spotlight on environmental grounds and the public will notice poor environmental practice. Therefore, it is important for good public relations to have proper regard for ecological issues on site.

The nature conservation bodies

This term is used here to represent the following organisations that have regional responsibility for promoting the conservation of wildlife and natural features:

- **Countryside Council for Wales**
- **English Nature**
- **Northern Ireland Environment and Heritage Service**
- **Scottish Natural Heritage**

Their remit includes protecting designated ecological sites, geological and geomorphological sites, and protected species. Within each area there are additional bodies that have an environmental remit, but no statutory process. These include county wildlife trusts and national and local environmental groups such as badger, bat and bird groups eg the RSPB, WWF and RIGS (Regionally Important Geological Sites) groups.

The main causes of problems

There are three ways in which ecological issues have to be considered on site.

Where species or areas of the site have been identified for particular protection

In this instance, studies will have been carried out to determine the best way of dealing with them. Site staff should be made aware of the special working methods that they should follow to protect the species or area of site.

Where protected species are discovered when the contractor is already on site and works have begun

Work should be stopped immediately, and the site manager should seek expert advice on how to proceed. Negotiations with the nature conservation bodies may then have to take place to discuss the best way forward.

To minimise general damage to ecology on site

Those working in construction should not only protect particular species and designated sites; it is important to take a responsible attitude to the natural environment as a whole. Be aware that most activities during construction can have a direct temporary or permanent impact upon the surrounding ecology, through such effects as:

- changes to water quality
- the destruction of places inhabited by plants and animals (this is a feature of most developments)
- interruptions to the movement of wildlife
- habitat fragmentation (for example by linear features such as access roads)
- vegetation damage through trampling by people or vehicles
- the removal of hedgerows and other vegetation
- dust (refer to 3.4)
- high noise levels disturbing adjacent ecology (refer to 3.3)
- changes in lighting
- damage, removal or burial of important rock formations or landforms.

Be aware that ecological resources off site may be affected by works on site.

Wildlife and natural features

How to avoid problems

The first step is to find out whether the client has identified any designated ecological sites or protected species; check that this information has been passed on to any contractors. Usually particular working practices to protect the ecological features will have been recommended; where these are given, follow the advice. If no advice is given, seek advice from head office or the person assigned with environmental responsibility in the site environmental plan (see Section 2.1.1).

Before work begins, identify and fence off any sensitive habitats and restrict the movement of workers to designated areas. This will help to minimise the damage that may be caused. For many categories of wildlife, for example nesting birds and roosting bats, the timing of work will be important, so correct scheduling may avoid problems. Liaise with the nature conservation bodies and with local environmental groups for advice.

Watch out on site

- Site staff are not expected to be ecological experts. However, they are expected to be reasonably aware of potential problems and to seek advice if necessary
- check for nesting birds, as it is an offence to interfere with them. If found, do not disturb them or cut down trees or shrubs. To avoid accidental disturbance to nesting birds do not fell or clear any trees or shrubs between March and July, if possible
- check whether any trees on site are covered by a tree preservation order. Talk to the local authority's tree officer.

Protecting trees on site

Damage to trees may be caused either through direct physical injury to the branches, trunk or roots or by changing the soil character around the roots. This can arise by various actions including: compaction (even one piece of machinery going over the roots can cause damage), raising soil levels, impervious covering around the tree roots, raising the water table, hazardous spillages, soil stripping, or excavations.

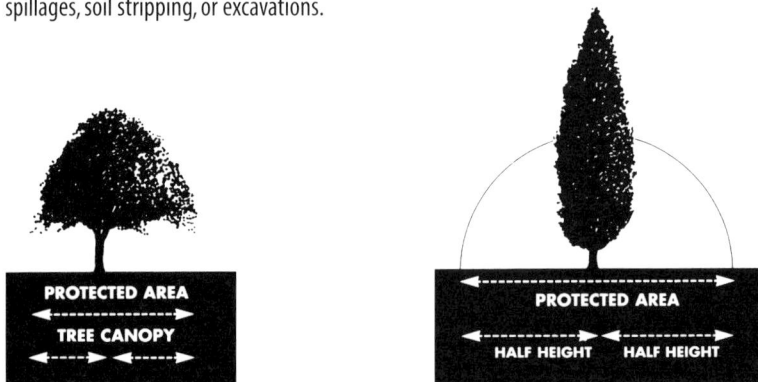

PROTECTED AREA
TREE CANOPY

PROTECTED AREA
HALF HEIGHT   HALF HEIGHT

To avoid damage to trees keep vehicles and plant away from them. The diagram opposite shows the area around a tree that needs protection. Put up temporary fencing to mark out this area. If it is essential to dig inside, dig only by hand. Do not cut or damage any roots greater than 25 mm in diameter within the protected area. When cutting a root, use a clean hand saw, not a spade or mechanical digger. Wrap damp sacking around any exposed roots until the hole is ready for backfilling. Backfill holes with care, to ensure that roots are not damaged, and compact the backfill lightly. Within the protected area, do not store spoil or building materials or drive vehicles. Keep toxic materials such as diesel and cement well away. Always avoid damaging bark or branches.

It is an offence to fell a tree that is covered by a tree preservation order without prior permission – check with the local authority. Be aware that there is a maximum amount of timber above which a felling licence is required from the Forestry Commission.

Works affecting a Site of Special Scientific Interest (SSSI)

Prior agreement with the nature conservation bodies is essential before carrying out works affecting an SSSI. It is an offence to carry out works on an SSSI without the permission of the nature conservation bodies. They usually require several months' notice before works begin. It is important to adhere to working methods once they have been agreed with the nature conservation bodies.

Working near water (refer to Section 4.20)

- Place a protective bund around ponds to prevent pollution of the water
- dewatering can affect the ecology of wetlands around the site. Consider monitoring water levels during the works.

Wildlife and natural features

Reinstating habitats

Measures to reinstate habitats damaged during the works may form part of the contract or planning conditions. Such measures have to be planned in advance and may require specialist advice. There may be particular requirements in selecting species for reinstating.

Dealing with key animals

Bats

If bats are likely to be disturbed, a bat survey will be required to establish the location and size of the roost. Advice must be sought on appropriate measures to move bats if required to allow works to continue. Contact the relevant nature conservation body, which may be able to inform you of a local bat group to liaise with.

Snakes and reptiles

Remember that all snakes and reptiles are protected and should not be killed or injured.

Amphibians

Great crested newts and natterjack toads are fully protected by law and to kill or injure them or to disturb their shelter is an offence.

Badgers

It is an offence to directly disturb a badger sett, or to carry out disturbing works close to a badger sett without a licence from the relevant nature conservation body. Guidelines on distances given are dependent on the activity. If a badger sett is discovered, stop work immediately and call in specialist advice.

Translocation of species

In some instances where habitat damage is anticipated and unavoidable, species may need be to be moved from site. This process is called translocation. It usually forms part of site preparation works and is carried out by specialists. Translocation can usually only be carried out within specific time periods (which depend on the species concerned), and therefore may influence the project programme. Both aftercare and monitoring of translocated species are essential. Where animals are found on site, or where translocation is considered, then various items of animal welfare legislation may apply (for example, the Protection of Animals Act 1911 and the Wild Mammals (Protection) Act 1996). Without prior planning and consideration, translocation may become a very expensive process.

Rocks and landforms

These are an important part of our natural heritage. Some rock exposures and landforms will be SSSIs (see Designated Sites section on page 93), whereas others may be locally important. If you are in doubt as to the importance of these features and how to protect them, contact the nature conservation bodies, which will give you contact details for your local RIGS (Regionally Important Geological Sites) group.

How to recognise when problems arise

As always, the best thing to do is to prevent problems arising. However, the following indicate that there are serious problems on site:

- bats or other nocturnal animals being seen during daylight hours or found on the ground near to the site
- injured birds/smashed eggs/young unaccompanied fledglings
- dead fish floating in watercourse affected by works.

Construction works can attract protected species. Sand stockpiles can become a nesting site for sand martins. Spoil heaps can be colonised by badgers, particularly if the surrounding water table is high. Be aware of this when moving materials.

What action to take if problems arise

Plants and animals

If you suspect that a protected plant or animal may be affected by your operations, stop and seek specialist advice before continuing.

If animals are injured contact the RSPCA.

Key Guidance

- *Bats in Buildings, Joint Nature Conservancy Council*
- *Great Crested Newts – guidelines for developers, English Nature*
- *Badgers – guidelines for developers, English Nature*

Wildlife and natural features

Legislation

Wildlife and Countryside Act 1981 (and amendments)

This provides for the establishment of Sites of Special Scientific Interest (SSSIs), which give protection to flora, fauna, geological or physiographical features of special interest. Some SSSIs are of international importance. These include Special Protection Areas (SPAs) for rare and migratory birds, established under the European Community Wild Bird Directive. There are also a number of wetland sites that are of international importance as waterfowl habits under the RAMSAR Convention of 1971.

Several hundred plant and animal species, for example the ghost orchid and spiked speedwell, and otters and bats, receive additional protection and must not be damaged under any circumstances. All birds are protected, and some, such as the kingfisher and barn owl, receive further protection and are covered by special penalties. Full lists can be obtained from the nature conservation bodies.

Mature trees in certain areas (especially urban) are protected by local protection orders and therefore must not be damaged.

Protection of Badgers Act 1992

Badgers are protected under this act. The aim of the legislation is to protect the welfare of badgers rather than conservation. The badger was too common to be included in the Wildlife and Countryside Act 1981.

Licensing

In certain cases it is possible to obtain special permission to disturb badgers and other wildlife; for example, the removal of protected species or their exclusion from sites where this serves a conservation purpose, or the interference with a badger sett for development. In the majority of these circumstances one of the nature conservation bodies will act as the licensing authority, but there are occasions when other organisations (eg Department of the Environment, Transport and the Regions) are the licensing bodies. Seek advice from the nature conservation bodies.

Other relevant legislation

● Hedgerow Act 1997.

Designated sites

Sites of international importance

RAMSAR *(UK is a signatory to this international convention)*	Sites listed under the Convention on Wetlands of International Importance. The protected sites are of importance as waterfowl habitats. *UK Statutory Designation:* SSSI (Site of Special Scientific Interest) *Contact:* the nature conservation bodies
Special Protection Area (SPA)	Classified under the European Community Directive on the Conservation of Wild Birds. *UK Statutory Designation:* SPA (all SPAs are SSSIs) *Contact:* the nature conservation bodies
Special Area of Conservation etc (SAC)	This legislation is based on the European Community Directive on the Conservation of Natural Habitats and of Wild Fauna and Flora (the Habitats Directive) *UK Statutory Designation:* SAC (all SACs are SSSIs) *Contact:* the nature conservation bodies

Sites of national importance

Site of Special Scientific Interest (SSSI)	Notified under Section 28 of the Wildlife and Countryside Act 1981. The sites have been studied in detail; particular aspects of the site will be highlighted in the site schedule. *UK Statutory Designation:* SSSI *Contact:* the nature conservation bodies
National Nature Reserve (NNR)	An NNR is managed primarily for nature conservation benefits. *UK Statutory Designation:* NNR *Contact:* the nature conservation bodies

Sites of regional or local importance

Local Nature Reserve (LNR)	A habitat of local significance that makes a contribution to nature conservation and to the opportunities for the public to enjoy wildlife. *UK Statutory Designation:* LNR *Contact:* designated by local authorities
Non-statutory nature reserves	These areas are established and managed by a variety of public and private bodies for specific ecological interest. *Contact:* county wildlife trusts, Royal Society for the Protection of Birds etc
Site of Importance for Nature Conservation/Key Wildlife Site	Usually adopted by the local authority for planning purposes. The name and status of this type of site varies considerably.

Archaeology

Why is archaeology important?

Archaeological remains are irreplaceable and are a valuable part of our national heritage. Encountering unexpected archaeological finds can affect both project programme and costs.

- Delays and costs can occur when works have to be stopped to allow for archaeological excavation (see Ancient Monuments and Archaeological Areas Act, page 98)
- damage caused to scheduled sites or monuments can result in prosecution and significant programme delays while the damage is assessed, and while scheduled monument consent is applied for to carry out repairs.

Sources of information on archaeology

Depending on the project's nature and location, work may have been carried out to investigate the archaeology of the site during the planning stage; this work is outlined here because it may be a useful source of information about the site. It is generally the client's responsibility to ensure that sufficient investigation has been undertaken to satisfy planning requirements (these are identified in local and national planning policies – see Key Guidance). However, this responsibility may transfer to the contractor in a design and build scheme.

Where nationally important archaeological remains (whether they are scheduled or not) are affected by a proposed development, the emphasis is on preserving them. If preservation in situ is not feasible, an archaeological excavation for the purposes of preservation by record may be an acceptable alternative.

When applying for planning permission, clients may be advised by the local planning authority to commission a desk-based archaeological assessment report as supporting documentation to the planning application. This report will review the archaeological potential of the site, assess the impact of the proposed scheme and propose a mitigation strategy.

If there is not enough information to decide how to proceed with the site, an evaluation (archaeological site investigation which may include trial pits/trenches, geophysical surveys, boreholes/auguring, air photography etc) may be proposed.

Using information from an assessment and/or evaluation, the local planning authority will decide how the archaeological issues on that site should be managed, and may then place a condition on the planning permission. They may issue a brief setting out what works are to be undertaken and then require a written scheme of investigation to be submitted for approval before works are commenced.

As archaeological evaluations and excavations usually occur early in the project they may be specified in the contract particularly where the contractor is to provide attendances for the archaeological work.

Managing archaeology on site

If it is likely that archaeological or historical features will be found during a project, the client will probably have commissioned some work on the site. Ask the client for the information they hold about the site.

Even if an investigation has been carried out there may still be a potential for unexpected finds to be uncovered during construction. The contractor's responsibilities and liabilities will depend on the particular contract and the site manager should be aware of these. The contractor is not expected to be an archaeological expert but must do the following:

- follow the contractual obligations, eg providing attendances and/or or access to professional archaeologists
- protect known archaeological and heritage sites
- report any significant finds arising during construction.

Comply with any contract and planning conditions

Identify any contractual obligations and conditions that may be attached to the planning permission (see above). Ask the client for any information that they hold on the site.

Protect known archaeological or historical features

Ensure that the proposed method of working complies with any obligations identified. Works that are located close to a site of archaeological or cultural significance can have a damaging impact. For example, vibration could cause cracking and subsidence in listed buildings; access roads could disturb historic areas.

The levels of vibration that can cause damage to buildings vary considerably. It is always worth agreeing the condition of susceptible buildings/monuments before works begin, and monitoring them thereafter. Seek specialist advice if vibration is likely to be a problem (see Section 3.3). Highlight the potential for significant effects on such buildings/ monuments to site staff and identify control measures before starting work on site.

Be aware that dewatering works can cause draw down of water from adjacent archaeological sites that may be well preserved because they are waterlogged. Dewatering may also cause differential settlement.

Archaeological finds will not necessarily delay works if surveys are properly planned.

Archaeology

Be prepared for unexpected finds

Whether or not known archaeological or historical features have been identified on your site, there is still the potential for unexpected finds to be uncovered during works. Materials to look out for during excavations include: burnt or blackened material, brick or tile fragments, coins, pottery or bone fragments, skeletons, timber joists or post holes, brick or stone foundations, infilled ditches.

If any finds are encountered

- stop work immediately in the area
- protect the find by fencing/blocking it off and contact the site manager
- consider whether it is worth seeking the site's own specialist archaeological advice on how to proceed. If addressed at the right time and in the right way finds may not necessarily affect the progress of the works. Time and cost may be saved if an archaeologist is called in as they may be available before the local planning authority archaeologist. An archaeologist employed by the company may be able to agree suitable mitigation strategies by telephone with the planning authority archaeologist. With the right advice the delay might be much less than any statutory period
- the site manager should contact the local archaeological officer at the local authority or county council, who may want to involve the heritage bodies (see below)
- it may be necessary to obtain separate Scheduled Ancient Monument consent before continuing work.

The heritage bodies

The heritage bodies are:

- **English Heritage**
- **Historic Scotland**
- **Northern Ireland Environment and Heritage Service**
- **Cadw (Welsh Historic Monuments).**

The heritage bodies have a general duty to conserving heritage, scheduling and undertaking research etc. Unless dealing with scheduled ancient monuments, contact the local planning authority archaeologist in the first instance for archaeological matters and the local authority conservation officer for listed buildings. The county/local authority/regional and island archaeologist has responsibility for planning issues.

Other information may be available from:

- British Archaeologists and Developers Liaison Group, c/o the British Property Federation.
- Code of Practice on the Treasure Act, Department for Culture, Media and Sport.

Be aware of the treasure trove rules (see below).

Removing a skeleton

Under the Burials Act 1857 it is necessary to obtain a licence from the Home Office to disinter any human burials. The licence is normally issued with conditions regarding the removal and disposal of such remains by an archaeologist.

Under the Disused Burial Grounds Act 1981 rather more stringent provisions relate to works affecting recognised cemeteries. A public notification procedure is involved as well as conditions being set down regarding the removal and disposal of human remains.

Treasure trove

Under the Treasure Act 1996 (for England, Wales and Northern Ireland), certain finds are treasure, including:

1. Objects other than coins, provided that they contain at least 10% of gold or silver and are at least 300 years old when found.

2. Coins, provided they are at least 300 years old when found.

3. Any object that is found in the same place as, or that had previously been together with, another object that is treasure.

4. Objects from the foreshore are not treasure, but may be the property of the landowner, normally the Crown Estate.

What to do if something that may be treasure is found

You must report all finds of treasure to the coroner for the district in which they are found either within 14 days after the day on which you made the find or within 14 days after the day on which you realised that the find might be treasure. You may report the find to the coroner in person, by letter, telephone or fax. The treasure is the property of the state; it has to compensate the freeholder of the land on which it was found.

Key Guidance

- *PPG16 Archaeology and Planning (November 1996) – this sets out current best practice for England and Wales (Department of the Environment, Transport and the Regions).*
- *NPPG5 Archaeology and Planning – current best practice for Scotland. Planning Advice Note 42 Archaeology, the Planning Process and Scheduled Monument Procedures for Scotland.*
- *Consult the heritage bodies.*

Archaeology

Legislation

Ancient Monuments and Archaeological Areas Act 1979

Any works to or within a Scheduled Ancient Monument and likely to damage that monument require the prior consent of the Secretary of State (referred to as Scheduled Monuments consent). Where consent is issued, it is frequently subject to conditions to prevent damage or to limit damage to agreed levels and with appropriate archaeological recording. Unauthorised works that damage a Scheduled Ancient Monument are a criminal offence, and significant penalties exist.

The Act also enables Areas of Archaeological Importance to be designated. Developers must then give six weeks "operations notice" to the planning authority of any proposals to disturb the ground, tip on it or flood it. A designated "investigating authority" has the power to enter the site and, if necessary, undertake archaeological excavations for up to four months and two weeks. After that time the investigating authority must cease excavation but can continue to enter the site to record and inspect the works.

Town and Country Planning Act 1990

Enables local authorities to protect a wide range of archaeological remains. Where development threatens to destroy remains, the authority can require appropriate investigation through a planning condition or legal agreement. In certain circumstances it can also secure the positive long-term management of sites.

The Planning (Listed Buildings and Conservation Areas) Act 1990

This provides for the compilation of lists of buildings of special, architectural or historic interest. Once a building has been listed, any object or structure fixed to the building is protected, as is any object or structure that forms part of the land and has done so since before 1 July 1948.

4 Construction processes

4.1	Introduction	100
4.2	Bored tunnelling	100
4.3	Brick/blockwork	101
4.4	Concrete batching	101
4.5	Concrete pours and aftercare	102
4.6	Demolition	103
4.7	Dredging	104
4.8	Earthworks	105
4.9	Excavation	106
4.10	Grouting	108
4.11	Microtunnelling	108
4.12	Piling (including temporary works)	109
4.13	Plant maintenance	110
4.14	Refurbishment of buildings	110
4.15	Repairs to exposed structural elements	111
4.16	Roadworks	111
4.17	Temporary works	112
4.18	Use of oils and chemicals	112
4.19	Use of small plant	115
4.20	Working near water	116
4.21	Working with groundwater	119

Construction processes

4.1 Introduction

At every stage of the construction process there is the potential for environmental problems to arise. This chapter identifies some of the main problems that may arise during a number of construction processes to highlight the areas that site staff need to be aware of and plan for.

Environmental issues are included here only when they are specific to a particular construction process. At all times, refer back to the guidance given in Chapter 3.

4.2 Bored tunnelling

Key issues	Refer to
Bored tunnelling may have **impacts on groundwater**, which could then have subsequent impacts on ecological habitats.	Water 3.1 Ecology 3.6 Wildlife and natural features 3.6
Plan the **disposal of spoil and slurry** arising from tunnelling ahead of works. Consider reuse options, but be aware of waste regulations. Minimise long-distance transport via road in order to minimise traffic impacts.	Waste 3.2 Key Guidance 1 below
Large-diameter tunnels near to ground level are most likely to cause **groundborne vibration**. Be aware of any sensitive buildings in the locality.	Noise and vibration 3.3
24-hour working may cause annoyance to neighbours near the tunnel portal.	Good public relations 2.2.2 Working hours 2.2.1 Noise 3.3
Contaminated ground or groundwater may be encountered during tunnelling. Develop a contingency plan for dealing with it. If it is encountered halt works immediately. Clear the site and ensure there is no smoking within 10 m of the site. Seek expert advice. Keep any contaminated spoil/groundwater separately from uncontaminated spoil/groundwater as it should be handled and disposed of appropriately. Tunnelling may cause a preferential pathway through which contaminants, mobilised by groundwater, may escape. Ensure that any contamination that is encountered is dealt with appropriately to prevent its spread.	Ground contamination 3.5 Water 3.1 Waste 3.2
Watch out for **unexpected archaeological finds** (this may include old cemeteries).	Archaeology 3.7
Identify listed buildings in the area surrounding the tunnel. Be aware that they may be susceptible to settlement.	Archaeology 3.7 Noise and vibration 3.3

Key Guidance

1. *Ground engineering spoil: good management practice, CIRIA Report 179, 1997*

4.3 Brick/blockwork

Key issues	Refer to
Bricks arising from demolition may have a reuse value or may be crushed and recycled as hardcore.	Waste 3.2
Avoid unnecessary wastage of materials by: • not over-ordering materials • storing bricks and blocks in their packaging to protect them and away from vehicle movements • preventing ready-mixed mortar from drying out • avoiding cutting and chasing.	Waste 3.2 Managing materials 2.2.4
Reuse **leftover excess bricks or blocks**, either on site or off site. If bricks or blocks are damaged, use them as hardcore on site access roads.	Waste 3.2
Segregate packaging wastes and return them to the supplier for recycling.	Waste 3.2
Dust from cutting and chasing may cause a nuisance.	See Dust, emissions and odours 3.4: Avoiding dust generation

4.4 Concrete batching

Key issues	Refer to
Locate concrete batching plant where they will cause least visual, noise and dust disturbance to neighbours. 24-hour batchers may cause light pollution.	The value of good public relations 2.2.2 Noise 3.3 Dust, emissions and odours 3.4
The **noise of motors and conveyors** annoys neighbours. Control the impact using screening.	Noise 3.3
Store aggregates onto an area of hardstanding to minimise wastage and avoid their contamination.	Waste 3.2 Managing materials 2.2.4
Additives are hazardous materials. Store drums properly to prevent spillage, and dispose of drums carefully.	Waste 3.2 Use of oils and chemicals 4.17
Control **surface drainage** from the area around the batching plant as it may be polluted. Obtain consent and dispose of appropriately.	Water 3.1
Accurately **predict the volume** needed to avoid over production and therefore wastage.	Waste 3.2
The **washout** from concrete mixing plant or cleaning ready-mix concrete lorries is contaminated with cement and thus highly alkaline; do not allow it to enter any watercourse. Try to reuse washout water as much as possible, then consider disposal options for on-site treatment and discharge (with consent) or off-site disposal.	Water 3.1: Disposing of water from site

Construction processes

If concreting operations are **near or in watercourses** (floating batchers), take special precautions to prevent the release of concrete.	Working near water 4.20 Water 3.1
Recycle returned concrete to reduce washout volumes and produce viable aggregate. This has been carried out by a number of major contractors on sites.	–
Encourage concrete wagon drivers to minimise the use of washdown water. Designate an area where washing down is permitted.	–
Cement dust is alkaline and harmful to ecology and will annoy neighbours. Follow the correct procedure for blowing cement into the silo, ensuring adequate supervision of the operation, the use of audible silo-mounted alarms and simple mass balance calculations.	Dust, emissions and odours 3.4 Wildlife and natural features 3.6
When **cleaning silos**, reverse-jet filters will minimise dust emissions by capturing and reusing cement dust.	–
Regularly **check pipelines and pumps** used to pump concrete to ensure that they are well maintained to prevent overfilling and spillage.	–
Maintain plant well to prevent noise and emissions.	Use of small plant 4.19

4.5 Concrete pours and aftercare

Key issues	Refer to
Control the storage, handling and disposal of **shutter oils**.	Use of oils and chemicals 4.18
Blowing dust and debris out of formwork can annoy neighbours. Consider suction.	Dust, odours and emissions 3.4 The value of good public relations 2.2.2
Consider the **wastes** that will be generated and plan their handling and disposal; for example: concrete curing compounds (spray applications), blackjack waterproofing (silane), PVC sheeting, frost protection materials.	Waste 3.2
A concrete pour may displace **wastewaters** from the hole. These may be contaminated with sediment and cleaning materials from the side of the structure. Dispose of them appropriately.	Water 3.1: Disposing of water from site
In or near a watercourse, control the placing of any wet concrete to minimise the risk of cement leaking into the watercourse. Shutter failure in such locations can cause major pollution in the watercourse.	Working near water 4.20

The **washout** from concrete mixing plant or cleaning ready-mix concrete lorries is contaminated with cement and thus highly alkaline; do not allow it to enter any watercourse. Try to reuse washout water as much as possible, then obtain consent and dispose of it to foul sewer via settling tank.	Water 3.1: Disposing of water from site
Large areas of concrete can create dust when they dry. Sweep them regularly.	Dust, emissions and odours 3.4
If batching concrete on site, minimise the environmental impacts.	Concrete batching 4.4

4.6 Demolition

Key issues

	Refer to
Before demolition begins, review the **disposal** options for the materials that will be generated. Reclaim and reuse materials where possible. Identify markets for materials. Segregate materials as they are generated. Dispose of any materials in accordance with your duty of care.	Waste 3.2 Key Guidance 1, 2, 3 below
If materials like concrete or masonry are to be crushed on site, check that you have obtained any necessary licences from the local environmental health officer (process authorisation under the Environmental Protection Act 1990).	Dust, emissions and odours 3.4
Before **removing or perforating tanks**, check that all of their contents and residues have been emptied by a competent operator for safe disposal. Pipes may contain significant quantities of oil or chemicals, and should be capped, or valves closed, to prevent spillage.	Key Guidance 4 Emergency response procedures 3.1
Noise and vibration annoy neighbours. Consider screening the works.	Noise and vibration 3.3 The value of good public relations 2.2.2
Dust from the demolition process may annoy neighbours and damage ecology near the site. Damp down structures during demolition.	Dust, emissions and odours 3.4
If **elephant chutes** are being used, ensure that each section is securely fixed, that the skip or lorry at the discharge end is covered, and that materials are dampened before being sent down the chute.	Dust, emissions and odours 3.4
Prevent dust escaping from materials in **lorries going off site**. If it is not possible to cover lorries leaving the site because there are pieces of material protruding, spray them with water just before they leave.	Dust, emissions and odours 3.4 Traffic and access routes 2.2.5

Construction processes

A contractor was fined £3000 for polluting the Cut at Bracknell, Berkshire, with gas oil. The incident arose when oil was drained from a compound where demolition work was being undertaken.

Key Guidance

1. *BS 6187 Code of practice for demolition, 1982*
2. *Waste minimisation construction – site guide, CIRIA Special Publication 133, 1997*
3. *Waste minimisation in construction – training pack, CIRIA Special Publication148, 1998*
4. *The reclaimed and recycled materials handbook, CIRIA publication C513, 1999*
5. *Remedial treatment for contaminated land Volume II: Decommissioning, decontamination and demolition, CIRIA Special Publication102, 1995*

4.7 Dredging

Key issues	Refer to
Plan the **disposal of dredgings** before starting works, to allow time for obtaining any permissions required under Waste Management Licensing Regulations 1994. Dredgings may be contaminated with substances such as oils and heavy metals; sample and test the sediments to provide vital information for considering disposal options. Aim to reuse the dredgings (for example, to improve agricultural land) rather than disposing of them to landfill.	Key Guidance 1
Disposing of dredgings at sea requires a MAFF licence. It can take up to three months to obtain a licence for a new site.	Ministry of Agriculture, Fisheries and Food
If **landfilling the dredgings**, obtain an exemption from landfill tax for dredgings arising from maintenance of navigable inland waterways	Waste 3.2: Landfill tax box
Obtain a discharge consent for returning **effluent from dewatering** dredgings to controlled waters.	Water 3.1
The dewatering sediments may give rise to an **odour** problem.	Dust, emissions and odours 3.4
Dredging may affect the **aquatic ecology**. Use an appropriate dredging technique to minimise the disturbance of sediment resulting in silting of the watercourse and potential mobilisation of contaminants. Consider using dredge mats and silt curtains. Disturbance of organic silts and dying weed may give rise to deoxygenation. Monitor and aerate if necessary.	Key Guidance 2, 3 Wildlife and natural features 3.6 Ground contamination 3.5
Working with **compacted sediments** may generate high levels of noise. Also, mechanical dewatering and compaction may lead to vibration.	Noise and vibration 3.3
General issues relating to **working near water** are shown in 4.20.	Working near water 4.19

Key Guidance

1. Guidance on the disposal of dredgings to land, CIRIA Report 157, 1996,

2. Inland dredging – guidance on good practice, CIRIA Report 169, 1997

3. The new rivers and wildlife handbook, The Royal Society for the Protection of Birds, 1994

4. British Waterways Environmental Code of Practice, British Waterways, 1996

4.8 Earthworks

This section considers the management of large movements of spoil on a project. For information on excavation, see 4.9.

Key issues	Refer to
Follow good practice in managing **stockpiles** of materials arising from earthworks.	See Checklist below
Minimise the arisings of surplus materials arising from earthworks by considering methods of improving the spoil (eg in-situ stabilisation).	Key Guidance 1
Plan the **disposal** of surplus materials arising from earthworks before starting works, to allow time for obtaining any permissions required under Waste Management Licensing Regulations 1994. Testing of the spoil may be required to provide information for considering disposal options. Aim to reuse spoil (eg for land profiling and raising) rather than disposing of it to landfill.	Key Guidance 1
If **contamination** is revealed during earthworks, halt works immediately. Clear the site immediately, ensure there is no smoking within 10 m of the site. Seek expert advice.	Ground contamination 3.5 Water 3.1: Emergency response procedures
Be aware of **unexpected archaeological finds**. Materials to look out for during excavations include: burnt or blackened material, brick or tile fragments, coins, pottery or bone fragments, skeletons, timber joints or post holes, brick or stone foundations, infilled ditches.	Archaeology 3.7
Surface water management. Keep water off unsurfaced areas using measures such as cut-off drains or french drains. Control and dispose of silty water properly.	Water 3.1: Discharging silty water
Earthmoving **plant and vehicles** used to transport materials from and around site may cause impacts from emissions, mud and noise. Construct appropriate haul roads. Maintain plant and vehicles. Use a wheel wash to minimise dirt on roads.	Noise and vibration 3.3 Dust, emissions and odours 3.4 Traffic and access routes 2.2.5 Plant maintenance 4.13 Use of oils and chemicals 4.18

Construction processes

Checklist – managing stockpiles

- *Store topsoil for reuse in piles less than 2 m high to prevent damage to the soil structure.*
- *Segregate different grades of soil.*
- *Position spoil and temporary stockpiles well away from watercourses and drainage systems.*
- *Minimise movements of materials in stockpiles to reduce degradation of the soil structure.*
- *Silty water formed by erosion of the stockpile must be managed correctly.
  See Water 3.1: Discharging silty water.*
- *Direct surface water away from the stockpiles to prevent erosion at the bottom.*
- *Place silt screens around spoil heaps to trap silt in any surface water runoff.*
- *Vegetate long-term stockpiles. This will prevent dust in dry weather conditions, and reduce erosion of the stockpile to form silty runoff.   Ensure adequate weed control.*

4.9   Excavation

Key issues	Refer to
Excavations may be formed within retained bases formed by sheet piling, diaphragm walling or secant piling.	Piling 4.12
Prevent **water** entering excavations. When water does enter excavations, take measures to avoid it becoming contaminated. Dispose of it properly.	See Checklist below Water 3.1
Be aware of **unexpected archaeological finds**. Materials to look out for during excavations include: burnt or blackened material, brick or tile fragments, coins, pottery or bone fragments, skeletons, timber joints or post holes, brick or stone foundations, infilled ditches.	Archaeology: 3.7
If excavation reveals **contamination**, halt digging immediately. Clear the site immediately, ensure there is no smoking within 10 m of the site. Where appropriate, try as far as possible to identify the extent and cause of contamination (eg prior landuse, spillage on site, rupture of subterranean pipeline) and halt any movement of contaminants. Seek expert advice.	Ground contamination 3.5 Water 3.1: Emergency response

If **asbestos** is uncovered unexpectedly during digging operations, halt digging operations at once and refill the excavation. Exposure of asbestos filings to the open air can result in widespread contamination as the particles are easily airborne far from site. Remove personnel immediately and secure the area. Contact site management immediately.	Contact site management.
Excavation **plant and vehicles** used to transport materials from and around site may cause impacts from emissions, mud, noise. Poorly maintained plant and vehicles cause more environmental effects than well-maintained plant. Use a wheel wash to minimise dirt on road.	Traffic and access routes 2.2.5 Plant maintenance 4.13 Use of oils and chemicals 4.18
Spoil arising from excavation can be recycled. Crush any rock arising and use on or off site. Store topsoil for reuse in piles less than 2 m high to prevent damage to the soil structure. Use excavated materials to form noise bunds and for landscaping – check whether planning permission is required.	Waste 3.2 Key Guidance 1

Checklist – dealing with water in excavations

- *Prevent water from entering excavations. Water running down the side of an exposed batter face may dislodge fine particles and take them into suspension. It may also cause collapse. Divert water by digging cut off ditches around the excavation or grading the ground.*

- *Prior to any excavation below the water table, including any site dewatering, inform the local Environmental Agency of the works to be conducted. (Refer to Section 4.20, Working near water.)*

- *If there is water in the excavation, do not allow plant or personnel to move about in it and stir particulate matter. Once particles are in suspension, particularly fine particles such as silt or clay, they can be difficult and expensive to remove. The quicker any free water (rain, seepage etc) is drained away the less opportunity to stir up and suspend particles. (Refer to Water 3.1 Discharging silty water.) Use the corner of the excavation as a sump and avoid disturbing that corner.*

- *Water in an excavation which is open for some time can be controlled by stone-filled edge drains leading to sumps.*

- *If groundwater is flowing into excavations, consider installing cut-off ditches, walls or wellpoint dewatering.*

- *Before discharging any water, always check that you have permission to do so and that the discharge complies with any conditions attached to that permission. (Refer to Water 3.1.)*

Key Guidance

1. *Ground engineering spoil: good management practice, CIRIA Report 179, 1997*

Construction processes

4.10 Grouting

Key issues	Refer to
Blowback from blockages or overfilling from pressure grouting with dry materials (eg cement) can cause significant dust problems. Working within an enclosure may be necessary in particularly sensitive areas, although health and safety precautions will be required for the workforce.	Dust, emissions and odours 3.4
Grouting in or near **contaminated ground** may displace polluted water from the excavation. Prevent the uncontrolled release of this water.	Water 3.1: Disposing of water from site (page 38) Ground contamination 3.5
Prevent the uncontrolled discharge of **cements and bentonite slurries**. Use a settlement tank to remove sediments and then obtain a discharge consent before releasing the effluent.	Water 3.1
Dealing with **waste grout**. Grout fines can be more successfully separated by the addition of a chemical flocculant, or by hydroclone separation or mechanical dewatering. This allows easier disposal of the constituents.	Waste 3.2 Water 3.1
Deal with any slurry waste (water mixed with silt) appropriately.	Water 3.1: Discharging silty water

4.11 Microtunnelling

Key issues	Refer to
Maintain **small plant** to minimise emissions.	Maintaining plant 4.13
Manage **wastes** arising from the works properly.	Waste 3.2
Noise from microtunnelling annoys neighbours.	Noise and vibration 3.3 The value of good public relations 2.2.2
Traffic entering and leaving site may disrupt normal traffic flow. Emissions from traffic will annoy neighbours.	Traffic and access routes 2.2.5 Dust, emissions and odours 3.4
Microtunnelling can damage **tree roots**.	Wildlife and natural features 3.6
Contaminated ground may be encountered during tunnelling. Develop a contingency plan for dealing with it. If it is encountered, halt works immediately. Clear the site and ensure there is no smoking within 10 m of the site. Seek expert advice. Keep any contaminated spoil separately from uncontaminated spoil, as it should be handled and disposed of appropriately. Tunnelling may cause a preferential pathway through which contaminants, mobilised by groundwater, may escape. Ensure that any contamination that is encountered is dealt with appropriately to prevent its spread.	Ground contamination 3.5

4.12 Piling (including temporary works)

Key issues	Refer to
Piling is an early activity – consider the effects of **gaining access in rural areas**. Consider using proprietary access track systems to minimise damage.	Traffic and access routes 2.2.5 Wildlife and natural features 3.6
Maintain plant regularly to maintain fuel efficiency.	Plant maintenance 4.13
Minimise the risk of spillage in **using oils and chemicals**.	Use of oils and chemicals 4.18
Noise and vibration will annoy neighbours. The noise levels created by piling vary with the method used (see table below). Some methods will not be allowed in urban areas, or other sensitive locations where the site has immediate residential neighbours; so use the right plant.	Noise and vibration 3.3 Key Guidance 1
Manage **bentonite** to prevent its release to the environment. Recycle it and then dispose of it properly.	Waste 3.2
Manage **wastes** arising from the piling operation itself. Wastes from bored piling may form a particular problem as the waste is often wet. Dispose of this waste properly.	Waste 3.2 Water 3.1
Piling **close to watercourses** forms a potential pollution risk.	Working near water 4.20
Contaminated ground may be encountered during piling. Develop a contingency plan for dealing with it. If it is encountered, halt works immediately. Clear the site and ensure there is no smoking within 10 m of the site. Seek expert advice. Keep any contaminated spoil separately from uncontaminated spoil as it should be handled and disposed of appropriately. Piling may cause a preferential pathway through which contaminants may escape. Ensure that any contamination encountered is dealt with appropriately to prevent its spread.	Ground contamination 3.5

Comparison of noise levels from types of piling	
Type of piling operation	**LAeq measured at 10 m**
Sheet piling using air hammer, no treatment	103–106
Acoustically treated hydraulic sheet piling	93
Impact bored pile	78
Continuous flight auger-injected piling	80
Vibroreplacement piling	82

Key Guidance

1. *BS 5228: Noise control on construction and open sites, Part 4 – Code of practice for noise and vibration control applicable to piling operations*

2. *How much noise do you make? A guide to assessing and managing noise on construction sites, CIRIA Project Report 70, 1999*

Construction processes

4.13 Plant maintenance

Key issues	Refer to
Designate an area within the site compound for routine plant maintenance. Surface this area and bund it. **Surface water runoff** may contain pollutants (eg oils), therefore control their release to controlled waters.	Water 3.1 Disposing of water from site
Ensure that appropriately **trained personnel** carry out repairs to plant.	–
Develop a protocol for disposing of **wastes** from maintenance. Recycle used oils. Dispose of old filters carefully as they contain substantial quantities of oil; some engine oils are special wastes. Recycle tyres. Try to find biodegradable oil substitutes.	Waste 3.2
Prepare for **spillages**. Display the site's emergency response procedure at the plant maintenance area. Ensure a spillage kit is kept there and that all personnel know how to use it. Carry a spillage kit in all repair vehicles.	Water 3.1: Emergency response procedure
Check replacement periods for **hydraulic pipework** and carry out regular checks. Use biodegradable oil for hydraulics. Blowouts regularly cause pollution incidents. Also check other pipework for signs of leaks; prevent plant dripping oil on the ground and polluting it.	–
Try to **minimise** the amount of refuelling or maintenance that occurs outside the designated area.	Use of oils and chemicals 4.18: Refuelling protocol
Install **drip trays** on plant where possible. Ensure that these are emptied regularly.	–
Store **hazardous materials** carefully to minimise the risk of spillage.	Managing materials 2.2.4 Use of oils and chemicals 4.18

4.14 Refurbishment of buildings

Key issues	Refer to
Explore the opportunities for using **reused and recycled materials** in the refurbishment. These materials may have arisen as wastes on site during the refurbishment or may originate off site.	Waste 3.2
Assess the **raw materials** that are proposed for use in the refurbishment. Check whether the use of such materials will affect the interior air quality.	–
There is a risk of exposing **asbestos** during the works, for example, by drilling into walls and ceilings. Ask the building owner and/or manager for detailed information on the location of asbestos in the building. Seek advice on what action to take if asbestos is uncovered.	–

Any **wastes** arising from the works must be handled appropriately and disposed of in accordance with the legislation.	Waste 3.2
Noise and vibration from works may annoy the building's inhabitants.	Noise and vibration 3.3

4.15 Repairs to exposed structural elements (bridge soffits, cladding etc)

Key issues	Refer to
Prevent **debris from works** falling onto the ground or water below. This should be incorporated into the working methods. It is especially important for works over public areas or near watercourses and sensitive ecological sites.	See Case Study below.
Store **hazardous materials** carefully to minimise the risk of spillage.	Managing materials 2.2.4 Use of oils and chemicals 4.18
Noise from the works may annoy neighbours.	Noise and vibration 3.3 The value of good public relations 2.2.2
If carrying out repairs over a watercourse.	Refer to Section 4.20

Case Study

Contractors shot-blasting a bridge used copper slag to remove old paintwork. This resulted in the production of a black grit (toxic to plants) falling down to the valley below. The grit settled along a 0.5 km stretch of the 90 m cliffs and had to be removed by specialists to prevent damage to a number of rare plants. The contractor was apparently unaware both that the site was a SSSI, and that there are other chemicals available for shot-blasting that are less harmful to the environment. The contractor should nevertheless have controlled the waste shot.

4.16 Roadworks

Key issues	Refer to
Minimise the risk of spillage in **using oils, bitumens and chemicals**.	Use of oils and chemicals 4.18
Noise and vibration may annoy neighbours. Night-time working may cause additional annoyance.	Noise and vibration 3.3 The value of good public relations 2.2.2
Traffic entering and leaving site may disrupt normal traffic flow. Emissions from traffic may annoy neighbours.	Traffic and access routes 2.2.5 Dust, emissions and odours 3.4

Construction processes

Much of the **waste material** arising from roadworks can be recycled.	Waste 3.2 Key Guidance 1, 2 and 3
Do not discharge **gully pot residues** to underground or surface waters. Alternative disposal options include discharging the liquors to the foul sewer (with the consent of the local sewerage undertaker), after solids settlement, or at a suitably licensed waste disposal site.	Water 3.1 Key Guidance 4

Key Guidance

1. *Use of recycling for road pavement construction and maintenance, County Surveyors' Society, 1994*

2. *Waste minimisation in construction – site guide, CIRIA Special Publication 133, 1997*

3. *Waste minimisation in construction – training pack, CIRIA Special Publication 148, 1998*

4. *The reclaimed and recycled materials handbook, CIRIA publication C513, 1999*

5. *Management of gully pots for improved runoff quality, CIRIA Report 183, 1998*

4.17 Temporary works

Key issues	Refer to
Take care when carrying out **temporary works near watercourses**.	Working near water 4.20
On-site fabrication of steelwork may be noisy and annoy neighbours.	Noise and vibration 3.3 The value of good public relations 2.2.2
For information on **piling**, see Section 4.12.	–
Reuse temporary works materials on site or off site.	Waste 3.2

4.18 Use of oils and chemicals

Key issues	Refer to
Avoid overordering of materials as this can lead to wastage of resources.	Waste 3.2
COSHH assessments need to be held on site for any potentially hazardous materials (Key Guidance 2). These provide advice on the type of **storage** needed for the chemicals, ie bunded areas, storage of flammable products in locked cupboards. Proper storage of hazardous materials reduces wastage and reduces the risk of spillages that could result in possible ground or groundwater contamination.	Key Guidance 1, 2, 3, 4 Ground contamination 3.5 Water 3.1 See information on bunding tanks below See information on storing oils and chemicals below

Use the correct quantity of chemicals; do not add more than required. Transfer chemicals between containers only within a suitably bunded area. Spillages outside this area could result in ground or water contamination.	Waste 3.2 Plant maintenance 4.13
In the **use of fuel**, follow the refuelling protocol to minimise the risk of spillage.	See the refuelling protocol below
Minimise accidental spillages but have **emergency procedures** in place in case of a spill. Store the product correctly. Ensure there is a spillage kit available and personnel trained to use it.	Water 3.1: Emergency response Key Guidance 1
Once the product is no longer required, store the product safely until it is needed again or find a market for the product. Any **disposal** of product or empty product containers should be in accordance with waste management legislation and the related COSHH sheet.	Waste 3.2 Key Guidance 1, 2, 3

Key Guidance

1. *Managing materials and components on site, CIRIA Special Publication 146, 1998*

2. *A guide to the control of substances hazardous to health in design and construction, CIRIA Report 125, 1993*

3. *Storage of packaged dangerous substances, HS(G) 71, HSE, HMSO, 1992*

4. *The storage of flammable liquids in containers, HS(G)51, HSE, HMSO, 1990*

5. *Environment Agency Pollution Prevention Guidelines:*
   – Safe Storage and Disposal of Used Oils
   – Above Ground Oil Storage Tanks
   – The Control of Spillages and Fire Fighting Run-off

6. *Construction of bunds for oil storage tanks, CIRIA Report 163, 1997*

Construction processes

Refuelling protocol
- Designate a bunded refuelling area, with an oil separator installed in the surface water drainage system.
- Avoid using remote fill points. Where these are unavoidable install suitable oil separators to the surface drainage system.
- Avoid refuelling close to watercourses. Where this is unavoidable, keep materials such as absorbent pads or booms readily available in case of spillage (Water 3.1).
- All refuelling must be supervised. Do not leave valves open unattended.
- Keep an emergency spill kit at each refuelling point. If mobile refuelling is carried out, ensure each bowser carries a spill kit (Water 3.1).
- Bowsers should have an automatic cut-out.
- Ensure that personnel carrying out refuelling are aware of the protocol and know what actions to take in an emergency.

Storing oils, fuels and chemicals
- Securely store all containers that contain potential pollutants (eg fuels, oils and chemicals).
- Clearly label containers so that appropriate remedial action in the event of a spillage.
- Regularly check taps and hoses for leakage.
- Avoid storing drums tightly against each other – store drums so that they can all be inspected for leaks.
- Prevent damage from vandalism (Site security 2.2.3). Ensure that all valves and trigger guns are vandal- and tamper-proof.
- Clearly mark the contents of any tank, and display a notice that demands that valves and trigger guns are locked when not in use.
- Store tanks or drums in a secure bunded container or compound that is locked when not in use.
- It may be necessary to have an impermeable base to any area where chemicals are stored in areas of particular groundwater risk. This should be identified in the contract but may be worth discussing with the Environmental Agencies.
- Provide separate fill pipes for each tank unless the tanks are interconnected by a balance pipe of greater flow capacity than the fill pipe.
- Mark fill pipes with the product type and a tank number where there is more than one tank.
- Before moving a drum, check the bung is secure.

> A contractor was fined £7000 for polluting Stocklough Beck with 22 gallons of diesel which had escaped from a generator left unattended during a refuelling operation.

Bunding tanks

- To avoid accidental spillage, bund tanks with a minimum capacity of 110% of the volume of the largest tank.
- Do not let bunded areas remain filled with rainwater or slops (ideally, provide a cover).
- Site tanks away from vehicle movements and mark them clearly so that they are visible and so that people know they are a potential risk.
- Do not put tanks where there is a direct link to surface watercourses or sewers.
- Avoid placing tanks on unmade ground, to reduce the risk of soil contamination.
- Protect from vandalism (Site security 2.2.3).
- The bund should be impermeable to the substance that is being stored in the tank.
- Position air vent pipes so that they can be seen easily and directed so that any discharge (eg in the event of the tank being overfilled) is directed into the bund.
- Fill points should be inside the bund.
- Fit any pumps sited outside the bund with a non-return/check valve installed in the feed line. (Also see Key Guidance 5 and 6.)

4.19 Use of small plant

Key issues	Refer to
Develop a **refuelling protocol** for your site and follow it.	Use of oils and chemicals 4.18
Maintain plant regularly to maximise fuel efficiency and prevent pollution incidents.	Plant maintenance 4.13
Prevent water pollution and ground contamination. Use drip trays under stationary plant to contain **oil leaks**.	Use of oils and chemicals 4.18 Water 3.1/Ground contamination 3.5
Ensure that only **trained personnel** use plant and that they use it for its intended purpose.	–
Remember that **water containing oils** or other chemical contamination cannot be discharged to watercourses or into or onto the ground.	Water 3.1
Secure plant from **vandals**, as they cause pollution incidents.	Site security 2.2.3
Noise and vibration from plant (eg generators and poker vibrators) can annoy neighbours and disturb ecology. Where possible, use quiet plant.	Noise and vibration 3.3 Wildlife and natural features 3.6

Construction processes

Ensure drip trays are used properly and emptied regularly.

4.20 Working near water

Examples of works near water include: channel diversions for culverts/bridges, sheet piling/scout piling for bridge foundations/groundworks, cofferdams, haul roads alongside rivers, discharges into rivers (see Water 3.1), dredging (see 4.7), bankside excavations, direct disturbance of watercourses, and overbridge cleaning and repairs (see 4.15).

Key issues	Refer to
Check that permission has been obtained for any **temporary works**.	See Temporary works box
For working from **pontoons and barges**, look at the checklist below.	See Checklist
Closely supervise all plant **refuelling**. Fill portable fuel tanks and spare fuel containers away from the water's edge and never overfilled.	Use of oils and chemicals 4.18
Keep adequate supplies of booms and oil-absorbent material available at all times for emergency use in case of a **spillage**. Dispose of any used oil absorbents properly.	Water 3.1: Emergency response
Prevent **pollution from plant** used near a watercourse. Maintain plant regularly. Use drip trays.	Plant maintenance 4.13
For information on **dredging**, see Section 4.7	–
Ensure that the runoff from **haul roads** near or over watercourses cannot enter the watercourse. If plant or vehicles have to be driven through a watercourse, erect temporary haul road bridges to prevent damage to stream beds and pollution.	Water 3.1

Where watercourse embankments are stripped of vegetation, stabilise them to prevent erosion. This may be done by placing biodegradable sheets down and seeding them using clover or fast-growing grasses.	Water 3.1
Do not wash tools out in watercourses.	
Prevent **dust or litter** blowing into watercourses.	Dust, emissions and odours 3.4 Waste 3.2
Be aware of potential direct or indirect disturbance to the bankside and in-stream **ecology**.	Wildlife and natural features 3.6
Ensure that works are secured from **vandals**.	Site security 2.2.3

When working over water consider using booms to control spillage.

Key Guidance

Environment Agency Pollution Prevention Guidelines:

- *General Guide to the Prevention of Pollution of Controlled Waters.*
- *Works In , Near or Liable to Affect Watercourses.*

Construction processes

Checklist – pontoons and barges

- Site all fuel tanks securely and safely on the vessel so that there is no chance of collision damage or accidental spillage overboard.
- It is an offence to discharge contaminated bilge water into any watercourse. If bilge water should become contaminated, it should be pumped to suitable facilities ashore or absorbents should be used (see Water 3.1: Disposing of water from site). Do not use detergents or emulsifiers in bilge water.
- Use hydraulic oil in floating plant.

Case Study

The Environment Agency has signed a contract with Texaco to provide a synthetic oil for hydraulic lubrication of all newly purchased mobile machinery, and a vegetable-based oil for all its chainsaws by 2005. All of the Agency's 200 diggers will use the synthetic hydraulic oil. Contractors working for the Agency will be expected to match its timetable for replacing mineral oil with biodegradable equivalents.

Temporary works

In England and Wales, the Environment Agency needs at least seven days' notice of any intention to temporarily or permanently divert the flow of a watercourse, carry out works over or within the river channel or commence operations in the river channel.

Obtain prior approval from the Environment Agency for all temporary works that involve construction, erection, re-erection or modification during work which:

- may interfere with the bed or banks or flood channel of any watercourse,
- is within 8 m of the bank of any main river, or
- is within 16 m of any tidal defence.

A subcontractor struck a sewer, and sewage was spilled into the River Medway. Both the contractor and the subcontractor were prosecuted. The contractor pleaded not guilty to charges for "causing polluting matter to enter the river" and knowingly permitting sewage to pollute the river and was found not guilty on the first charge, but guilty of the second, for which it was fined £5000. The subcontractor was fined £4000 on each of the charges.

4.21 Working with groundwater

Key issues	Refer to
Notify the relevant Environmental Agencies where extensive dewatering is to occur so they may issue a conservation notice, ie guidelines for dewatering to prevent the alteration of the groundwater regime and/or contamination of groundwater.	Water 3.1
Any **groundwater abstracted** from a site needs to be disposed of. It is generally accepted that the least environmental effect is caused if this is returned to groundwater. Seek advice from the Environmental Agencies before implementing this solution, because a discharge consent may be required.	Water 3.1
There is a risk of **mobilising ground contamination** when working with groundwater, and **a potential risk of ground instability**. If any contamination is suspected, seek specialist advice before proceeding. Before starting works, investigate the level of these risks and apply appropriate control measures. Consider monitoring for petrol/oil or diesel compounds that will float on the groundwater surface and move with the flow of water.	Ground contamination 3.5
Works affecting groundwater may have an impact on adjacent ecology. Monitor water levels around the site. Dewatering may cause a change in groundwater levels and affect river/stream flows (see case study below). A solution is to monitor water levels in sensitive areas and recharge with the extracted groundwater. However ensure that the recharge water is of the correct quality and temperature – a settling lagoon in the loop can assist in this. Seek specialist ecological advice before attempting such solutions.	Wildlife and natural features 3.6 Water 3.1
Minimise the risk of spillage in using oils and chemicals.	Use of oils and chemicals 4.18

Newbury Bypass Case Study – monitoring

While the contractor was dewatering along the line of the road, water levels were monitored in the adjacent SSSI. If the water level fell below acceptable levels, the groundwater was recharged using water that had been held for some time in a settlement tank. This ensured that the water was returned to the groundwater at an appropriate temperature.